C.B.W.R.Co.
Ex. "H"
John Hodson
Mar 1, 73
Lorenzo D Smith
Oct. 28 91
Fannie Bickford
May 27 83
Chas. Van Zile
Mar 16. 07
R. Painter
Wm W. Pribble
Feb. 5. 1903
John Kruse
Feb 3. 83
J.C. Haynes
Jan 3. 83
Alice M. Norton
Apr 22. 07

SEEING THE FOREST *for the* TREES

MENASHA CORPORATION
AND ITS
ONE HUNDRED YEAR HISTORY
IN COOS BAY, OREGON

WILLIAM A. LANSING

SEEING THE FOREST *for the* TREES

MENASHA CORPORATION
AND ITS
ONE HUNDRED YEAR HISTORY
IN COOS BAY, OREGON

1905 to 2005

WILLIAM A. LANSING

The Monroe Press
Eugene, Oregon

dedication

It is with a great sense of personal gratitude that I thank the descendants of the Company founder for giving me the chance to unfold my wings and continue the entrenched corporate tradition of believing that people matter; as coined by the motto of the Menasha Forest Products Corporation: "People First, Product Second".

Printed and bound in Portland, Oregon, U.S.A.

The Monroe Press
Eugene, Oregon

ISBN: 0-9767649-0-3

First Printing: May 2005

Publisher/Editor Tom Hager
Designer Chris Michel

Cover photograph of skid road courtesy Oregon Historical Society 7908
Back cover photograph courtesy Oregon Historical Society 7729
Title page photograph of timber fallers at Cunningham Creek, Coos County, 1911, courtesy of Coos County Historical Museum
Table of contents photograph of C. R. Smith inspecting Menasha logs

All photographs courtesy of the author unless otherwise noted.

table of contents

foreward

Historical literature in its most basic format is a chronological listing of dates and facts; in a more elegant form, it combines the facts of an era with the stories of the people who brought it to life.

One Third Crew, One Third Boat, One Third Luck, *written to celebrate Menasha's 125th anniversary*

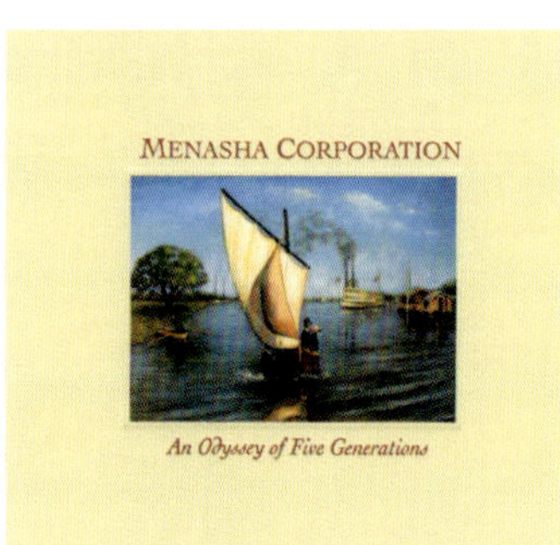

Menasha Corporation: An Odyssey of Five Generations *written to celebrate its 150th anniversary*

The history of a corporation is fleeting, as the records are often thrown away as no longer useful or, from a legal perspective, potentially damaging. Further, the people who moved the organization forward drifted into retirement and are gone.

Such is the case with Menasha Wooden Ware Company as in 2005 it celebrates its 155 years of existence in Wisconsin under the descendents of the Smith family patriarch and its 100 years of operations in Coos Bay, Oregon. In rummaging through old Menasha corporate documents, I regretted that I had not asked those who preceded me for their remembrances. Fortunately, there were sufficient documents, notes, and letters in the files in Wisconsin and Oregon to piece together the story of Menasha out West.

Two books have been written about Menasha Corporation and its predecessor companies: the first written by Mowry Smith (and Giles Clark) on the occasion of the 125th anniversary of the Company entitled *One Third Crew, One Third Boat, One Third Luck*. The second book about Menasha is entitled *Menasha Corporation An Odyssey of Five Generations* and was written in 1999 by Richard Blodgett for the 150th anniversary of the company. By necessity, both books dealt with the depth and breadth of Menasha in summary fashion only. To cover 150 years of business experiences could not be done in any other way.

As Menasha approaches the 100th Anniversary in the Pacific Northwest, this book *Seeing the Forest for the Trees: Menasha Corporation and Its One Hundred Year History in Coos Bay, Oregon, 1905–2005* will focus on the history of the Menasha Wooden Ware Company (and its successor companies) as it moved from Menasha, Wisconsin, to Coos Bay, Oregon. The book will discuss the geopolitical environment under which Charles Robinson Smith, President of Menasha Wooden Ware Company (1899–1916), and other company executives, staked their claims to timberland ownership in Saskatchewan, Idaho, Oregon, and Washington and to a lesser extent in New Mexico, Utah, and British Columbia.

The reader will be given a glimpse into actions of the federal government as it encouraged the country to "move West." The Homestead Act, the Donation Land Claim Act, the Oregon and California Rail Road Grants, the Stone Act, and the Military Wagon Road Grant lands all played a role in Menasha's history. It would eventually be the strong hand of the government in Washington D.C. that would bring an end to the land subsidies and the often unscrupulous processes used to secure timberland in the West.

Charles R. Smith would compete with other "Timber Barons" as they all sought to replace the dwindling timber supplies that supported their mills in the Midwest. Charles A. Smith (no relation to the Smiths of Menasha) and Frederick Weyerhaeuser, as well as many others, will be found in the text.

C. R. Smith as a young man

This book leaves behind the town of Menasha, Wisconsin, and the company's many successful manufacturing ventures during its 155-year history. It will focus specifically on the timber connection of Menasha's history out West and its related forest products manufacturing operations. The final chapter introduces the new Menasha Forest Products Corporation that was formed in 2001 to manage the company's timberland assets west of the Rockies.

The reader will find many subsidiary names scattered throughout the book as acquisitions and new construction projects by Menasha Wooden Ware Company were isolated into separate business names. It would not be until Richard L. Johnson became president of Menasha in 1961 that the corporate structures were consolidated under one name–Menasha Corporation.

While I have endeavored to develop the book in a chronological fashion, there are really no distinct events that open or close each era. Each chapter does however possess a common theme. I hope the logic will be evident.

Bill Lansing
March 2005

menasha timeline

Utter City and the Henryville Mine on Isthmus Slough (Courtesy of Gordon Ross)

The Simpson shipyard and sawmill in North Bend, with the four-masted schooner Admiral at dock (Coos Art Museum, Victor C. West Collection)

1827
Elisha Dickenson Smith born on March 29th in Brattleboro, Vermont, the eldest son of seven children

1850
Elisha Dickenson Smith and Julia Ann Mowry married October 24th
Donation Land Claim Act passed by Congress

1851
John Preston initiated cadastral land survey system

1852
Pail factory in Menasha, Wisconsin purchased by Elisha Smith for $1,200
Congress named Salem capital of Oregon Territory
Gold discovered on the beach in southern Oregon

1853
Congress carved the Washington Territory out of the Oregon Territory
John C. Tolman acquired the future town site of Marshfield

1854
First coal shipment from Coos Bay
Coos County created on January 19th

1855
Rogue River Indian War began
Henry Heaton Luse purchased sawmill in Empire City from Robert Moore

1856
Asa Meade Simpson built a sawmill on the bank of Coos Bay at a site later to be known as Old Town North Bend

1857
Asa Simpson built first ship *Arago* in his North Bend shipyard

1859
Congress granted Oregon statehood on February 14th

1862
The Homestead Act passed by Congress

1863
Thomas Winchester purchased 53 acres of land, later to become the town site of Yarrow (North Bend) for $66.81

1864
Military Wagon Road Grant Act passed by Congress

1866
The Oregon and California Railroad land grant issued (4,220,00 acres)

1869
Congress authorized the Coos Bay Military Wagon Road land grant

1870
The Coos Bay Wagon Road Company organized (grant for 105,120 acres)

1872
Oregon and California Railroad completed the line to Roseburg
Henry Heaton Luse acquired a major portion of the Marshfield townsite and adjacent tidelands

1873
Empire City declared a port of entry by the United States and a customs house established

1875
Menasha Wooden Ware Company incorporated on May 24th
Henry Hewitt Sr. elected president; Elisha Smith employed as general superintendent

1878
The Timber and Stone Act passed by Congress

1881
Elisha D. Smith elected president of Menasha Wooden Ware Company,
Charles Robinson Smith elected treasurer, and Henry S. Smith elected secretary

1883
First meeting of the subscribers of the Oregon Southern Improvement Company in New Bedford, Massachusetts
Oregon Southern Improvement Company purchased 7,000 acres of timberland and farmland from Henry Heaton Luse, along with the Luse sawmill in Empire City and development lots in Empire City and Marshfield

1884
Oregon Southern Improvement Company purchased lands owned by the Coos Bay Wagon Road Company (94,000 acres) for $800,000
Oregon Southern Improvement Company purchased franchise of the Oregon Southern Railway and Transportation Company for $5,000
Elijah Smith (no relation to the Menasha family) elected president of Oregon Southern Improvement Company
Large sawmill built by the Oregon Southern Improvement Company in Empire City for $150,000

C. R. Smith at his home

1887
Assets of the Oregon Southern Improvement Company purchased at a sheriff's sale by William J. Rotch and William W. Crapo for $1,516,666.66, and they immediately formed the Southern Oregon Company

1888
The Southern Oregon Company acquired the lands of William J. Rotch and William J. Crapo for $1,500,000

1890
Town of Yarrow (North Bend) platted on June 17th

1899
Elisha Dickenson Smith died July 7

1901
Julia Ann (Mowry) Smith died

menasha timeline

Libby Coal Mine narrow gauge train (Coos County Historical Museum)

Seeley and Anderson railroad trestle collapse 1912 (Oregon Historical Society 36254)

1902
Crater Lake National Park opened
Louis Simpson acquired townsite of Yarrow and replatted it as North Bend
Menasha purchased 7,700 acres from the State of Idaho for $85,561
Weyerhaeuser purchased 31,000 acres in the Millicoma forest from Northern Pacific Railroad for $5 per acre
North Western Wooden Ware Company formed by George P. Wright, et.al.

1903
City of North Bend is incorporated
Menasha purchased a large tract of land in Idaho from the Northern Pacific Railroad for $54,488

1904
Kruse and Banks established their shipbuilding company on land where the large Weyerhaeuser mill complex was later built in 1951
Menasha established an office in Coeur d'Alene, Idaho

1905
Menasha acquired 225 square miles (144,000 acres) of timberland from Herman Finger along the Saskatchewan River in Saskatchewan, Canada
Oregon land fraud trials pursue wrongdoers
Menasha purchased the first timberland parcel (318 acres) in Coos County, Oregon, from Henry B. Taylor of Waupaca County, Wisconsin, for $2,200

1907
Siuslaw National Forest created by executive order of President Theodore Roosevelt

1908
Southern Oregon Company acquired by Menasha along with the Coos Bay Wagon Road lands in Coos County, Oregon
Charles Axel Smith built the big mill at Bunker Hill in the upper Coos Bay
The hopper dredge "Oregon" began deepening the channel in Coos Bay from the entrance to the mouth of Isthmus Slough

1909
Oregon Caves National Monument created

1910
Tacoma Eastern Railroad line built to Morton, Washington

1911
Oregon Department of Forestry created

1912
Elijah Smith, president of the Oregon Southern Improvement Company and the Southern Oregon Company, died in Portland, Oregon, April 11th

1913
Brookings Lumber Company established the town of Brookings
Menasha acquired 50 percent interest in timberlands in New Mexico from John Garvin and William MacFarlane for $93,750
Menasha acquired 160 acres in Utah

1914
Congress revested Coos Bay Wagon Road land grant
Menasha acquired timber license in British Columbia

1915
First timberland purchased by Menasha in the State of Washington

1916
Charles R. Smith died. F. D. Lake became the new president of Menasha Wooden Ware Company
The Southern Pacific Railroad completed line from Eugene down the Willamette Valley to Coos Bay

1917
West Fork Logging Company fire in Lewis County, Washington

1918
Thomas Kearney, an attorney from Racine, Wisconsin, and business consultant, became president of Menasha Wooden Ware Company
Menasha acquired another small tract of timberland in Utah

1919
Menasha consented to the revestment of the Coos Bay Wagon Road lands back to the federal government
Menasha sold interest in New Mexico timberlands for $9,289

1921
Willis Miner elected president of Menasha Wooden Ware Company

1922
North Western Wooden Ware Company in Tacoma, Washington, purchased
A fire on Front Street in Marshfield destroyed a major portion of the old business district.

1923
Highway 38 along the lower Umpqua River opened for auto traffic

1924
Clarke-McNary Act aided federal-state forest fire protection

1926
Menasha reformed as two companies; the Menasha Wooden Ware Company as the holding company and Menasha Wooden Ware Corporation as the operating company Willis H. Miner, president of both

1927
Empire Development Company incorporated by William G. Robinson and Louis J. Simpson
Finger Lumber Company in Saskatchewan sold to Winton Brothers of Minneapolis, Minnesota

1928
South jetty on the Coos Bay bar completed

1929
Menasha built first Wood Fibre plant in Tacoma, Washington, to handle the waste from the North Western Wooden Ware factory

The Simpson Lumber Company rail line on Daniels Creek. Charles Bickford in cab of locomotive (Coos County Historical Museum)

Rock barges pushed by tug Tonquin *down the Coos River on way to Coos Bay jetty construction (Coos County Historical Museum)*

menasha timeline

Long log truck on way to mill (Oregon Historical Society 16864)

Timber falling by hand, 1950s

1933
Tillamook Fire destroyed 240,000 acres of forest

1935
Congress authorized Bonneville Dam, which was completed in 1938

1936
Carlton Smith elected president of Menasha Wooden Ware Company and Mowry Smith Sr. elected president of the Menasha Wooden Ware Corporation
McCullough Bridge, a Public Works Administration project on Coos Bay, dedicated on June 6th, completing Highway 101 (the Oregon Coast Highway)
Town of Bandon nearly totally destroyed by fire; eleven people killed

1942
Incendiary bombs dropped on Mt. Emily in Curry County, Oregon by a small plane launched from a Japanese submarine off the southern Oregon coast

1944
Marshfield changed its name to Coos Bay

1945
Tillamook Fire area "re-burn" destroyed 180,000 acres
"Balloon bombs" launched into the jetstream by Japan, dropped incendiary devices on Coos County and the western United States

1946
Coos Head Timber Company organized by Willis Smith and Chuck Duecy

1949
Menasha formed partnership with Coos Head Timber Company to build Douglas-fir plywood plant in North Bend, Oregon. The authorized capital for the new Menasha-Coos Head Plywood Corporation was $800,000
State Department of Forestry began replanting Tillamook Burn

1950
Menasha bought out Coos Head's interest in the Menasha Coos Head Plywood Corporation and formed the Menasha Plywood Corporation

1951
Weyerhaeuser opened a large saw mill in North Bend

1952
Carleton L. Smith died

1954
Major logging activity began on Menasha timberlands in Washington
Menasha acquired Irwin and Lyons timberlands (21,667 acres in South Fork Coos River drainage) along with extensive logging equipment, the Coos River Boom Company, and a logging camp at Tioga Creek for $3,468,348

1955
Highway 101 relocated between Coos Bay and Bandon
Wood Fibre plant built North Bend, Oregon
Menasha's Washington timberlands "mothballed"

1956
Menasha acquired Al Peirce timberlands in Coos County, Oregon (11,948 acres) for $2,000,000
Menasha acquired Ball Lumber Company sawmill and timberlands (1,882 acres) at Hauser, Oregon, for $450,000
The upper splash dam on the South Fork Coos River burned
Menasha experimented with new forestry project on its Davis Slough ownership

1957
Ball Lumber Company mill destroyed by fire

1959
Menasha sold "Menasha Hill" (Telegraph Hill) in Coos Bay to Vernon Scott of Eugene

1961
Mowry Smith Sr. elected chairman and Richard L. Johnson elected president of Menasha Wooden Ware Corporation, and Carlton R. Smith elected chairman and Mowry Smith Jr. elected president of Menasha Wooden Ware Company
Southwestern Oregon Community College established
Menasha built North Bend paperboard mill

1962
Menasha Wooden Ware Corporation changed its name to Menasha Corporation
Columbus Day Storm hit Coos County and western Oregon
Bunkhouse at the Menasha Tioga logging camp destroyed by fire—no injuries

1963
Wood Fibre plant built at Albany, Oregon
Menasha plywood plant in North Bend, Oregon, expanded and new office built at 1515 Sheridan Street

1964
Mowry Smith Sr. died July 4th

1967
Wood Fibre plant built Grants Pass, Oregon
Menasha plywood plant in North Bend permanently closed on September 1st

1969
Doyle Milling Company veneer plant acquired at Norway, Oregon
Wood flour plant built Grants Pass, Oregon
Menasha closed its company logging operations
Menasha timberland ownership in Oregon and Washington totaled 84,971 acres

1970
Donald C. Shepard Sr. died
Menasha started timber sale program from fee lands to local mills in Oregon
Intensive forestry programs initiated on Menasha timberlands

Logs behind the Upper Dam on the Coos River awaiting release, summer 1955

Timber fallers, 1950s

menasha timeline

Donald C. (Tad) Shepard and President George Bush

Rosevelt Elk on Menasha land

1971
Donald G. Turner died

1972
Congress established The Oregon Dunes National Recreation Area

1973
Menasha restarted its Washington tree farm forestry and timber sale programs

1974
South Slough Estuarine Sanctuary established
City of Lakeside incorporated
The Bay Area Hospital opened its doors on May 19th in Coos Bay
Wood Fibre Division built plywood glue extender plant in Centralia, Washington
Menasha purchased Hiller 12-E helicopter

1980
Menasha timberland ownership in Oregon and Washington totaled 94,971 acres
Mt. St. Helens erupted

1981
North Bend paperboard mill sold to Weyerhaeuser
Menasha timber sale program halted
International log sales program started with the People's Republic of China
First export vessel was the *Javara*, with documentation work done by Caffall Bros. Forest Products Company in Wilsonville, Oregon
Donald C. Shepard became President and CEO of Menasha Corporation

1982
Mowry Smith Jr. retired as senior vice president; William A. (Bill) Lansing elected vice president for Menasha's timber operations including Wood Fibre Division

1984
Menasha helicopter crashed near Scofield Creek, Douglas County, Oregon—no fatalities
Menasha purchased a Hughes 500-C model helicopter
Menasha started non-fee timber purchase program in Washington
Menasha started Timberland Assistance Program
Menasha Development Corporation formed to sell non-strategic lands in Pacific Northwest

1987
Menasha purchased plywood glue extender plant in Marysville, Washington, from Asbury Graphite for $600,000
Menasha formed partnership with Caffall Bros. Forest Products to build CB Cedar sawmill in Coos Bay, Oregon. Construction cost $3.5 million

1989
Weyerhaeuser Company closed its large North Bend mill
Weyerhaeuser purchased CB Cedar sawmill

1990
Hughes 500-C model helicopter traded for a Hughes 500-E model

1993
Menasha opened a timber office in Lewiston, Idaho, to purchase timber and logs for the Japanese market

1995
Coquille Indian Tribe opened The Mill Casino at the former Weyerhaeuser plywood mill site in North Bend, Oregon

1997
Wood Fibre Division of Menasha sold to Ace Chemical of Vancouver, British Columbia

1999
Freighter *New Carissa* is grounded off the north spit of Coos Bay

2000
Idaho timberland office closed
Menasha timberland ownership in Oregon, Washington, California, and Idaho totaled 104,000 acres

2001
Menasha Forest Products Corporation formed as a stand-alone company on April 1st
Thomas Prosser, Jim Sansick, William A. Lansing appointed as Transition Board of Directors

2004
Menasha Forest Products Corporation converted to Real Estate Investment Trust (REIT) on January 1st
Menasha exited the non-fee timber business
Coquille Indian Tribe purchased all land once occupied by the Weyerhaeuser sawmill and chip operation in North Bend

2005
Current Directors of Menasha Forest Products Corporation:
James C. Janning, Chair
Curt N. Smith
Lydia B. Smith
John Wilkinson
William Shepard
William A. Lansing, President and Chief Executive Officer

Current Officers of Menasha Forest Products Corporation:
William A. Lansing, President and Chief Executive Officer
Thomas E. Jackman, Vice President, Chief Financial Officer and Secretary
Scott Starkey, Vice President
Katherine Barton, Treasurer
Martha King, Assistant Secretary

Riparian zone left tp protect stream traversing Menasha forest land in Oregon

Coos County old-growth (Courtesy of Warren Dick)

chapter one

C. R. Smith Comes to Coos Bay

Charles Robinson Smith was looking at his company's future. The twentieth century had just begun, and Smith—everyone called him "C. R."—was touring the coast of Oregon looking for raw materials for his company's woodenware factories. He saw Oregon's Coast Range, mountains thick with massive spruce, hemlock, and Douglas-fir. He saw rivers big enough to float the huge logs to sawmills.

At some point in his travels he saw the final piece of the puzzle: a natural harbor on the Pacific, close to the timber, big enough for large ships, and surrounded by enough level land to accommodate the mills, yards, and docks he intended to build. It was called Coos Bay.

C. R. was a man of vision, an archetypal Midwest American industrialist known for his wit, drive, boundless energy, and big plans. He, along with his brother Henry, had inherited the Menasha Wooden Ware Company—a leading Midwest maker of wooden barrels, pails, tubs, and kegs—from their parents, Elisha Dickinson Smith and Julia Ann Mowry, a strong-willed and hardworking couple who founded and built the firm at Menasha, Wisconsin, during the last half of the nineteenth century.

C. R., a Princeton graduate whose classmates included Woodrow Wilson, took active control of the company in 1899 at age forty-four and immediately began plans for expansion. Growth could not happen on the scale he wanted without access to new timber. The great forests of Minnesota, Michigan, and Wisconsin were being cut fast, and his company would soon run out of the wood needed to feed the nation's growing demand.

So C. R. looked west.

In 1903 he purchased land and timber rights in Idaho. In 1905 he bought the first of many tracts of timberland in Coos County, Oregon, and acquired 225 square miles (144,000 acres) of timberland along Canada's Saskatchewan River from Herman Finger, forming the Finger Lumber Company, Ltd. Substantial investments were made in the Saskatchewan area, including building a sawmill and planing mill. Finger Lumber was reported to have had a standing timber inventory in excess of 600 million board feet growing on their property.

In 1911 C. R. acquired the Southern Oregon Company, a major forest products company in Coos County, Oregon, with its substantial timberlands, along with their big sawmill in Empire City, a tugboat, a rail line, and a couple of stores. The timber from these purchases was originally acquired to supply a major portion of the lumber needs of Menasha as C. R. began contemplating expanding his woodenware factories out west.

On his trips to the Pacific Northwest C. R. undoubtedly found traveling difficult as he reached the hinterlands. Ocean beaches often provided the only efficient public roadway along the rugged Oregon and Washington coasts and travelers were required to use ferries to cross the frequent river estuaries.

Charles Robinson (C. R.) Smith President of Menasha Wooden Ware Company 1899–1916

C. R. Smith inspecting Menasha logs stored in a lake in the Midwest (Courtesy of Lydia Smith)

Elisha D. Smith and Julia Ann Mowry Smith

THE BIRTH OF MENASHA

The story of the current Menasha Forest Products Corporation began 155 years ago with a strong-willed couple who wanted to make their own way in this developing nation. The New England aristocracy at the time probably viewed them as a bit rebellious as they married and made their way west from Rhode Island to Menasha, Wisconsin—leaving their families behind. Their names were Elisha Dickinson Smith and Julia Ann Mowry.

Mowry Smith, Jr. (a fourth-generation descendent of the Menasha patriarch) probably said it best in the beginning pages of his book, "One-Third Boat, One-Third Crew, and One-Third Luck", written to commemorate the 125th anniversary of the Menasha Woodenware Company in 1974:

> Elisha fortunately chose a mate who shared his pioneering spirit and his courage to face new frontiers. Their long journey to Menasha not only constituted a challenge to these qualities but also taught the young couple something of rigors of transportation of that day. They traveled by stagecoach, sailing vessel, steamer and railroad. From Woonsocket (Rhode Island) to Buffalo they went by rail car; from Buffalo to Detroit by steamboat, then again by rail across what is now Michigan to the eastern shore of Lake Michigan; then by steamer to Chicago and on to Milwaukee and Sheboygan (Wisconsin); at Sheboygan by stage to Fond du Lac then on to Oshkosh (at the south end of lake Winnebago) then boat to Menasha.

The newlyweds arrived in Menasha, Wisconsin, in 1850, and made it their home for the next fifty years. It was here that they raised their family. Elisha built a large woodenware business manufacturing tubs, pails, and barrels supported by four sawmills in Auburndale, Stevens Point, Wrightstown, and Apple Creek, Wisconsin. Along the way he acquired over 100,000 acres of timberland in northern Wisconsin, Minnesota, and Michigan to support the business. His enterprise became known as the Menasha Wooden Ware Company.

Throughout his life, not only did Elisha tenaciously manage the affairs of the Menasha Wooden Ware Company for almost a half-century, but he also became well known in the central Wisconsin area for his philanthropy.

Elisha Smith died on July 7, 1899, and Julia Ann (Mowry) Smith died in 1901.

Elisha D. Smith

A Menasha stock certificate granted to Donald C. (Tad) Shepard Jr., a great-grandson of Elisha Smith

It was worth the difficulty. C. R. must have been awed by the massive Douglas-fir, Sitka spruce, red cedar, and hemlock trees found growing on the steep hills of the Coast Range in Oregon and Washington. Of interest, yet unrelated to C. R.'s investment decision, was an interesting phenomenon surrounding the forests of the Coos Bay region: The trees in the back-country were all about the same age.

How could this be? Arthur V. Smyth, a lifelong career forester for Weyerhaeuser, in his book *Millicoma, a Biography of a Pacific Northwestern Forest*, describes in some detail how the Weyerhaeuser Company tree farm just east of Coos Bay came to pass. While he does not claim that the eighteenth-century explorers saw a mighty forest fire scorching the hills of the Coos and Millicoma Rivers in 1765, he does document the work by company foresters in 1945 and 1946 who inventoried some 200,000 acres in both river drainages. He writes, "When the results were analyzed, tree after tree was close to the same age—180 years—meaning something big had happened somewhere around 1765 and that something big was a fire. In fact, the foresters in 1945 noticed the charred scars in the scattered ancient survivors of that fire."

In 1915, a horse-drawn wagon was used to assist an automobile across Ten Mile Creek at Lakeside, Oregon

While Smyth makes his points about the origin of the forests of the Coos Rivers, one can only speculate as to the intensity of the conflagration. Smyth states, "Two or three hundred years of forest growth went up in flames in probably not much more than twenty-four hours." Dow Beckham, an Oregon historian and author of several books about the industries around Coos Bay, noted in a personal interview with Bob Corthell that the historical records of Coos Bay indicate that about a foot of ash washed down and covered the mudflats of the Bay, killing all the native oysters.

Fires (caused by lightning or Native Americans) have been for centuries an integral part of the ecology of the Douglas-fir forests of Oregon and Washington. There is ample evidence that prior to the discovery of the New World, Native Americans continually modified the forest composition of the Pacific Northwest with fire to enhance their food opportunities.

The 500,000-acre Biscuit Fire in Curry County, Oregon, 2002 (Coos Forest Protective Association)

In addition to the fire of 1765, which created today's Coos River basin forests, many other wildfires left tell-tale scars on the surviving trees: The fire of 1868 that ravaged the Oregon Coast from Port Orford to the Umpqua River, the Nestucca fire of 1845, the Yaquina fire of 1846, and the Bandon fire of 1936. All of these events had an impact of the Oregon forests we see today and from which innumerable benefits are provided.

Indeed, if it were not for the fire ecology of the Pacific Northwest, the political and emotional controverseries surrounding the "virgin stands of old-growth Douglas-fir" would not exist. Instead, the hillsides would be covered with dense thickets of hemlock and red cedar.

It is interesting to connect this fire history with later events. Elijah Smith (no relation to the Smiths of Menasha), future President of the Southern Oregon Company, in testimony given before the Oregon Senate in 1888, claimed many acres on his property in the South Coos River basin were worthless as they were covered with nothing more than charred snags!

The author, as a young forester helping to manage the Menasha forests in the Coos River watershed, quickly learned the various colloquial terms of the old timers. Red fir referred to the forests that originated roughly about the time of the 1765 fire (or about 200 years of age in 1969) and yellow fir, or "bastard growth," referred to the trees that survived the fire of 1765 and were mostly rotten and fire-scarred, leaving only a portion of the tree that was of much value.

Table Rock Fire, Douglas County, Oregon, 2002 (Coos Forest Protective Association)

Coal seam on Kentuck Slough on the east side of Coos Bar (ca. 1904). Lorenz Dow Kinney on left (Coos County Historical Museum)

The Newport Mine, which came to be known as the Libby Mine, was for many years the most productive in the Coos Bay area. The nearby Southport Mine also produced extensive volumes of coal. Both mines lay beneath current Menasha timberlands just south of the city of Coos Bay (Courtesy of Dow Beckham)

Two coal bunkers at the mouth of Coal Bank Slough are visible in this photograph from the early 1900s. The property known later as Menasha Hill is visible above the bunker at the upper left of the photo (University of Oregon)

Coos Bay coal bunker (Coos County Historical Museum)

Once the coal was pulled from the mine, it would be loaded into small rail cars for the journey to Coos Bay. (Coos County Historical Museum)

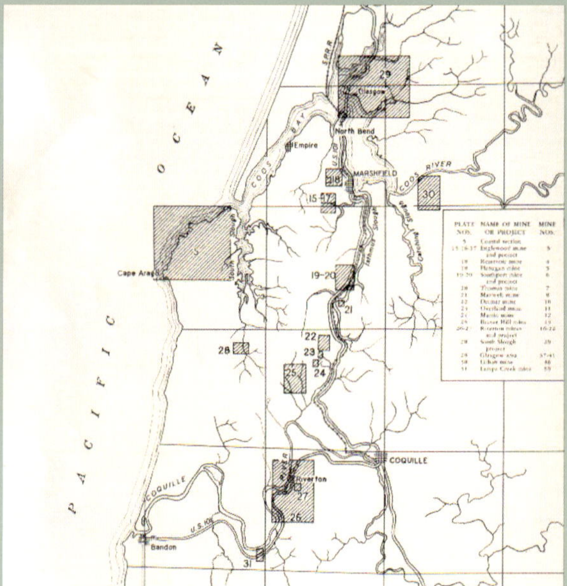

Coal mines located on Menasha timberlands surrounding Isthmus Slough, south of Coos Bay (Baldwin)

COAL ON THE BAY

Most people living in the Coos Bay region today would have a difficult time imagining just how extensive the coal mining operations were during the turn of the century. In fact, Coos Bay would not have grown as it did without its ties to "black gold."

The Gold Rush in California created a coal rush as well, when the growth of cities like San Francisco led to a huge demand for fuel. The closest major coal source, it turned out, was around Coos Bay. By the 1890s, more than twenty coal-carrying ships per month crossed the treacherous bar at the mouth of Coos Bay bearing cargo bound for California. Much of the land immediately south of the Bay contained extensive seams of coal. In some spots, the coal seam reached to the surface and chunks of coal could be picked up and burned in the stove. This is even true in some spots today.

Several coal companies sprang up to tackle the challenges of extracting the valuable fuel, and by the fall of 1855 the coal boom was on. Several of the major coal mines along Isthmus Slough were on timberlands later acquired by the Menasha in 1956. In total there were fourteen operating coal mines on or near these future Menasha timberlands. Freeman Lockhart, one of the first developers of a coal mine on the Bay, shipped thirty tons of coal the first year he started business, in 1854. Another discovery on the Donation Land Claim of Perry Marple (the land was later owned by a Menasha subsidiary) near Empire City produced a full shipment of coal to San Francisco in that same year.

As with other underground mining operations during the late 1800s, animal power was used to haul the carts full of coal from deep within the earth to the daylight. The carts traveled along crudely constructed narrow-gauge rail lines.

After the carts were pulled to a small storage bunker and dumped, larger rail cars were used to transport of the coal some five to eight miles to the large storage bunkers at tidewater.

Many of the mine owners lacked experience in coal, and they struggled with the lack of transportation as regular service of ocean steamers or schooners had yet to be established in the harbor. Further complicating the mining process were the coal formations themselves. The coal seams in the Coos Bay area were severely fractured by faulting, dipping deep into the landscape and being so close to the Bay were constantly plagued by seeping water. This made shaft mining the coal particularly challenging.

Eventually, a combination of lower demand and increasing difficulty in finding and mining the coal led to the end of the coal era. After 1923, no Coos County coal was shipped from Coos Bay.

Second-growth timber referred to trees that were around 100-150 years of age and seeded naturally into the clearings created by the fires or followed the natural evolution of forest types in the area. Third-growth forests were composed of trees between forty and sixty years old, originating again from a natural seeding from the surrounding older Douglas-fir trees.

Even today, wildfires, although generally contained to less than a few hundred acres on private industrial timberlands, are a constant threat during the dry summer months. While extensive forest road systems provide quick access for "initial attack," the 500,000-acre Biscuit Fire in southwestern Oregon in 2002 originated within the roadless Kalmiopsis wilderness area managed by the USFS—and given the immense size of the fire, it took a major change in the weather and the eventual winter rains to extinguish the flames.

An old-growth red cedar tree in Coos County, Oregon (ca. 1900) (Coos County Historical Museum)

Along with so many of the early timber pioneers of the West, C. R. Smith probably thought he was looking at a never-ending supply of timber from these vast forests for the wood needs of a growing nation and his hungry mills.

But he needed more than trees. He needed the ability to mill the logs and ship his products around the world. That was when he decided to center his company's western expansion in Coos Bay.

The towns surrounding Coos Bay area were already thriving when C. R. arrived. Like many early settlements out west, the initial growth along the southern Oregon coast had been driven by the hunt for gold. Gold was found on Jackson Creek near the town of Jacksonville, Oregon, in 1852, and soon was discovered on the beaches at the mouth of the Coquille River, some twenty miles south of Coos Bay. It took the miners until the summer of 1853 to reach the coast and begin prospecting in the tributary streams.

An example of a pit saw operation in New England. This same technique was used in the first days of creating lumber in Coos County in the early 1850s (Yale University)

The local miners, with their "get rich quick" attitudes, often caused hostile relations with the local Native American tribes—especially those in the Rogue River drainage region. The tribes around Coos Bay, however, did not participate in the Indian-white conflicts of 1855-56.

Gold would not "pan out" as a viable industry for the south coast, but the region had now been discovered and explorers from the fledgling communities along the Umpqua estuary to the north and the Rogue River valley to the south traveled to Coos Bay to examine its prospects.

After the yellow gold fever died away, energy needs for the rapidly expanding town of San Francisco created a new opportunity for local entrepreneurs around Coos Bay. They turned their attention to "black gold" obtainable from mining the Coos Bay-area coal field, as well as lumbering from the immense trees standing at the edge of tidewater. Both of these enterprises required access to ocean shipping from the large and protected Coos Bay harbor. Coos Bay was a natural location to supply the wood and coal needs of the booming San Francisco area. In the end, all these abundant resources would lead to the rapid and extensive development of coalmines, farming, sawmills, and ship-building in the area.

Red Dog pit sawmill in Douglas County, Oregon (Douglas County Historical Museum)

As C. R. began acquiring and harvesting lands in Coos County, a new and difficult challenge awaited: how to physically find his property lines in the field, where few roads existed.

The government survey in Oregon began around 1850 and continued for the next fifty-plus years. The rugged individuals who undertook this work had only basic survey instruments with which to work and, without much in the way of access, were often out in the back country for weeks on end. They were called the Cadastral Surveyors. It was their job to set the township and

Bull team skidding a large old-growth Douglas fir log in the late 1800s (University of Oregon)

section corners for each 640-ace block—for all of Oregon. They often used simple wooden stakes to mark corners but, thankfully for those who followed some years later, referenced these corner stakes to nearby trees (called bearing trees). Menasha surveyors still use these original corner-points 125 years later to establish property boundaries

Smith's approach to acquiring land was to purchase lands from individual homesteaders who had purchased their properties from the federal government in 160-acre tracts. Obviously, more surveying was needed to define his ownership on the ground before any harvesting could begin. He turned to the Coos County Surveyors for help.

Logging in the early days around Coos Bay, however, was limited by the topography and the strength of animals used to drag the logs from the stump to the water's edge or to a nearby mill. Once a log was delivered to the water it could be floated to the mills that were scattered along the Bay. While steam engines were quickly being developed to replace other methods of mechanical power in mills and railroads, they were slow to be adapted to logging operations. It would not be until the late 1880s that the first steam donkeys, as they were affectionately called, reached the logging business. The first one arrived in Coos County about the same time C. R. did, in 1904.

As the original pioneers in Coos County began carving their homesteads out of the forests, rudimentary sawmill techniques were imported from other parts of the United States—all using human power. The first area "sawmill" was a pit saw operation in the Coquille valley region.

Getting the logs to a central site was one problem. The next was sawing them into boards. When C. R. first visited Coos Bay, he found several local businessmen already operating steam-powered sawmills in the area. Prominent among them were Henry Luse and master shipbuilder Asa Meade Simpson (who used his mill to cut lumber for his ships as well as for export). Luse and Simpson were great competitors, each constructing a sawmill in 1856, Luse in Empire and Simpson in Old Town North Bend (as that section of the current town of North Bend was then known). The Simpson mill was built at the southeasterly "toe" of the current Simpson Heights area about three-quarters of a mile up the Bay from Highway 101's McCullough Bridge, very near the current location of the Menasha Forest Products Corporation headquarters in North Bend.

Ten-man survey team including A. N. Gould as Coos County surveyor, 1914 (Douglas County Historical Museum)

Along with his sawmill and logging operations around Empire City, Luse also ran cattle in the sand dunes on the north side of the Bay—reaching his ranch only by boat. Locally, this property was known as the Henderson Ranch, and it would play an important role for Menasha in the region some 130 years later.

While the general record indicates that Simpson and Luse built their mills in 1856, a report in a master's thesis (author unknown but probably done at the University of Oregon) indicates:

In May 1853, Asa M. Simpson visited the area and decided that it could be developed easily. This California

ship-owner and sawmill operator, originally from Maine, noted the wealth of timber around the bay and decided to establish a sawmill there. His equipment arrived in 1856, but the ship carrying it was wrecked while entering the bay and one of Simpson's brothers lost his life in the mishap. Owing to this delay, the first sawmill in the area was built by Henry Luse at Empire City.

Nathan Douthit, a Coos County historian, indicated that the first sawmill in Coos County was built by George Wasson and a partner near Bullards (south of Coos Bay) in 1853.

C. R., in addition to finding lumber mills on the Bay, also found a few factories making finished wood products. For instance Asa Simpson's son Louis, along with various local businessmen, had organized the North Bend Manufacturing Company in March 1903 to make Port Orford cedar doors (at the rate of 300 per day), window sashes, porch columns, moldings, and other millwork material. Louis also helped to organize the North Bend Shingle Company in 1904. By 1907 North Bend Shingle was cutting 100,000 shingles per day, but on August 24, 1911, the mill was destroyed by fire. It was rebuilt only to burn again on June 16, 1918, and was never rebuilt.

After looking the area over, appraising the timber, exploring land values, assessing the shipping possibilities, and eyeing his competitors, C. R. Smith made his decision. On September 21, 1905, he purchased Menasha's first parcel of timberland in the Coos Bay area: 318 acres in Coos County sold by Henry B. Taylor of Iola, Waupaca County, Wisconsin. Smith's company paid $2,200 for the property. Three months later he bought another 480 acres from the same people for $5,760. These were the first steps in creating what is today the Menasha Forest Products Corporation.

The work of C. R. Smith, Asa and Louis Simpson, E. B. Dean, Luse, and other local mill owners and lumbermen offset the loss of coal shipments from Coos County; lumber would more than

Henry Heaton Luse (Coos County Historical Museum)

Asa Meade Simpson (Coos County Historical Museum)

Henry Luse Store and Dock at Empire, Oregon (Oregon Historical Society 86529)

Henry Heaton Luse Mill (Coos County Historical Museum)

make up for the decline in "black gold." By the mid-1960s, Coos Bay carried the title of the "World's Largest Forest Products Exporting Port." It was a title earned in great part by the vision and courage of people like C. R. Smith and the growth of his Menasha businesses in the region.

The modern city names of "Coos Bay" and "North Bend" were not in common use at the time. Before 1902 North Bend was known as Yarrow, and before 1944 what we now call the city of Coos Bay was Marshfield.

James C. Tolman (Stephen D. Beckham)

The old names figure prominently in local history. In 1853 James C. Tolman (one of 19 members of the Coose [sic] Bay Company, later to become the Surveyor General of the General Land Office) came to the area to acquire property as a personal investment. He took possession of a tract of land later to become the town site of Marshfield and built a log house on the hill land. In 1854, two men named Williams and Crosby entered into a contract with Tolman to build a store and warehouse to bring in goods to supply the growing town. For inducement, Tolman gave each man two lots in the town site. At this point, the name of Marshfield was chosen in memory of a similarly named town in Massachusetts.

In 1854, Tolman granted one-half interest in Marshfield to A. J. Davis. Due to the declining health of Tolman's wife, Tolman moved to Jackson County. Apparently Davis did not live in the area and he hired Wilkins Warwick to represent him. Warwick is reported to have lived in the Tolman log house until 1856, when he left for Iowa.

The North Bend Mill and Lumber Company (ca. 1915) (Coos Art Museum, Victor C. West Collection)

Two years before he left, Warwick, on behalf of Davis, deposited enough money ($1.25 per acre) in the General Land Office in Roseburg to qualify for ownership of the townsite of Marshfield under the requirements of the Donation Land Claim (DLC) law of 1850.

Unfortunately for Warwick (and Davis) the DLC stated that "Donations hereafter to be surveyed in Oregon and Washington Territories shall in no case include a town site or lands settled upon for the purposes of business or trade and not for agriculture." In 1857 the US land survey was extended to Marshfield and the Warwick Donation Land Claim (as it was known) was designated by the General Land Office.

By 1864, the town of Marshfield had grown to some three hundred residents, many of whom sought title to the land underlying their homes. Since no patent had been issued to Warwick/Davis, and apparently many of the residents were aware of the fraudulent nature of the original claim, the locals hoped they could have the Warwick DLC cancelled and their own claims honored. In addition, some inhabitants who had built on pilings below the high-tide mark tried to perfect their ownership claim under the Oregon's Tidelands Act of 1872.

The Nann Smith, *named after C. A.'s daughter, built at the Simpson shipyard in North Bend, carrying lumber from the C. A. Smith sawmill in Marshfield (Coos Art Museum, Victor C. West Collection)*

Just prior to all these problems, Davis sold out to John McCombe and Henry H. Luse.

Local government officials approved the claims of the individual owners, but Luse appealed to the Secretary of the Interior and sought assistance from one of Oregon's Senators, George H. Williams. After Williams became Attorney General for Oregon in 1874, the contest was decided in favor of the Warwick title—and consequently Luse—on some 185 acres in Marshfield.

Since there was no deed on record for the Warwick claim, Henry Luse sent his attorney to Iowa to obtain a quitclaim deed from an Isaac Dickinson who had purchased some of the property. The quitclaim deed was received for $12,500. However, court challenges continued under the townsites exclusion section of the original DLC Act.

Luse even filed ejection papers on some of the inhabitants of "his land." At this point the litigants became despondent, as it seemed that Mr. Luse's "unrelenting hand" was upon them. Luse finally sold his claim in Marshfield to the Oregon Southern Improvement Company (OSIC)—later to become part of the Menasha holdings.

In a letter written July 4, 1884, from William Granger, OSIC's manager in Coos Bay, to Prosper W. Smith, attorney and treasurer of OSIC in New Bedford, Massachusetts, Granger refers to numerous lawsuits between OSIC and the people of Marshfield regarding the title to tidelands fronting the town. Granger goes on to report that: "Mr. Luse sued the parties then occupying them for ejection—not on the grounds that he owned the land, but that it was obtained from the government through misrepresentation—that it should have been deeded to him (by him paying for it), as his lands bordered on these tidelands."

Since Luse was now out of the picture, Granger met with the individual owners and told them that since the OSIC was now the owner of the uplands, the company would rather settle their claims than pursue it in court. The OSIC board followed Granger's advice and settled the title to these private lots for a "small sum of money."

The Henry Luse land in Marshfield included the best residential property in town. After the Southern Oregon Company (successor of OSIC) was acquired by Menasha in 1911, there were 230 lots left on what later became known as "Menasha Hill." Evidently Menasha did not sell these lots and the city "grew around them," elevating the future value of each lot. Today this area is known as "Telegraph Hill," after the famous location in San Francisco.

North Bend, Oregon (ca. 1898) (Coos County Historical Museum)

In March 1959, Menasha agreed to sell twenty-three acres of Menasha Hill to Vernon Scott of Scott-Built Homes of Eugene, Oregon. Scott planned to build twenty "upper-bracket" homes on the site. Since some of his homes would be multi-family structures, rezoning would be required and Scott made this a prerequisite for the purchase from Menasha.

In late April 1959, the Coos Bay Planning Commission recommended rezoning a portion of the purchase for the multiple residential classification of up to forty apartment units, however the Coos Bay City Council rejected the rezone plat on June 9, 1959, with Scott stating, "In view of the city council's refusal of my preliminary plat and rezoning proposal, I intend to go ahead and exercise my option to buy the Menasha Hill property. I will skin off all the timber and sell it. . . . I haven't decided yet whether to build $9,000 houses or something more expensive."

North Bend Shingle Company (ca. 1905) (Coos Art Museum, Victor C. West Collection)

Today, Telegraph Hill has both a grouping of multi-family apartment houses and some very "expensive" single-family homes.

As a sidebar, the author recalls an experience involving a strip of land on the Menasha Hill property. About 1972, Menasha received a $4,000 bill for back sewer taxes on property within the city limits of Coos Bay. If not promptly paid, the city would foreclose. The author, along with a Menasha surveyor, went to the site and determined that the culprit property was a strip of land some twenty-five feet in width and 100 feet long. The problem was that this particular parcel of land ran directly through a beautiful garden of very old well-tended patch of rhododendron plants.

The property had been part of the original Luse land acquired by the Oregon Southern Improvement Company. The Luse deed provided for a catchall legal phrase, "and all other land within the city limits of Coos Bay." When the property was later sold, the catchall phrase was omitted and the strip of rhododendron legally belonged to Menasha, not the owner of the

The North Bend Manufacturing Company (ca. 1905) (Coos Art Museum, Victor C. West Collection)

Simpson Shipyard North Bend (Oregon Historical Society 105314)

Marshfield, Oregon (ca. 1907) (Oregon Historical Society 10461)

rhododendrons. The landowner felt otherwise and refused to pay the sewer lien, as did Menasha. The property remains in limbo to this day (2005)!

Today's North Bend was also known by a different name before 1902: It was called Yarrow. As with Marshfield's evolution into Coos Bay, the change of Yarrow's name involves some important local history.

On May 5, 1863, a man by the name of Thomas D. Winchester journeyed to Roseburg (the US Land Office) and purchased from the United States Government a 53.45-acre tract of land for $66.81—a price of $1.25 per acre, the going rate for land at that time. This land was located where the current city of North Bend is located. Eleven months later Winchester sold the land to Charles Merchant for $500; and eleven months later Merchant sold the land to Charles Eckoff for $400. Eckoff kept the property for twenty-four years until December 30, 1889, at which time he sold it to Thomas Symons for $16,000. On June 17, 1890, Symons filed papers in Portland, Oregon, to plat the property for the proposed city of Yarrow.

On June 27, 1902, Louis J. Simpson (son of Asa Meade Simpson), L. F. Falkenstein, and W. U. Douglas traveled across country to Erie County, New York, to meet with Symons and his wife, where they negotiated the sale of the Yarrow property for $25,000. The original "deal" was a contract, but by October 22, 1902, Simpson had paid the entire amount in full, and Simpson along with his wife Cassie filed a new plat in Coos County calling the city "North Bend."

Louis Simpson would play an ever-increasing role in the growth of North Bend until he filed for bankruptcy on August 15, 1940.

Charles Axel "C. A." Smith

Charles Robinson was not the only Smith prominent in Coos Bay during the early part of the twentieth century. Charles Axel (C. A.) Smith—no relation to C. R.—built a big mill on Isthmus Slough in 1908, applying pressure to the federal government to improve the shipping channel in Coos Bay so that lumber shipments could increase. C. A. was a Swede who came to the United States in 1866, settling in Minnesota with his family, where he built one of the largest sawmills in the region. As with other timber men of the upper Midwest, Smith's timber supply dried up and he moved to Oregon.

C. A. Smith rebuilt the E. B. Dean mill after it was destroyed by fire in 1910 (Coos County Historical Museum)

In December 1906, C. A. purchased the E. B. Dean sawmill located on the east side of Isthmus Slough on Coos Bay, 10,000 acres of land containing around 350 million board feet of standing timber, and one mile of Bay frontage at Marshfield.

He also built a sawmill at the foot of Bunker Hill on the Bay that went into operation in 1908. After financial difficulties, C. A.'s mill was recapitalized and became the Coos Bay Lumber Company—which would in turn become Georgia-Pacific. C. A. was cutting big timber that required big equipment and a solid foundation to the buildings as the big logs banged through the mill. Some of the band saws he used were in excess of sixty feet in circumference.

The C. A. Smith sawmill in the upper part of the Bay was the cornerstone for leveraging the federal government to improve the shipping channel and bar crossing into Coos Bay (University of Oregon)

As the other Smith in town—Menasha's C. R. Smith—planned his company's future in Coos Bay, he saw two potentially limiting problems. One centered on the Bay itself, which was in places too shallow for the big vessels needed for lumber shipping and required crossing a sometimes-dangerous bar to reach the sea. The other was acquisition of the huge tracts of timberland needed for efficient logging. Solving these issues would determine the success of Menasha in the coming century.

Tugs were used to tow sailing vessels out of Coos Bay and across the shallow bar to the Pacific Ocean (University of Oregon)

chapter two

Land, Water, and the Big Mill

The entrance into Coos Bay from the Pacific proved quite a challenge to early captains trying to navigate their way in to load cargoes of coal or lumber.

The trials in crossing the bar at Coos Bay were many: shifting channels, lack of steam tugs for pilot-assist service, failing winds at crucial moments when sailing ships were in perilous waters, incomplete information about the entrance, and lack of lights and markers to assist mariners in lining up their vessel within the channel. Many ships were lost in attempting to cross the bar.

The first reconnaissance of the Coos Bay bar by the US Coast Survey (subsequently known as the US Coast and Geodetic Survey) was done by James Lawson in 1861. Lawson's efforts were plagued by fog, rough seas, and lack of steam-powered boats. He departed for San Francisco where, on March 21, 1862, he reviewed his efforts in a letter on the Coos Bay project:

> *I cannot furnish my hydrological work in as satisfactory a style as I desire. This is the first time I ever attempted such work, and that little experience has taught me many things. . . the work outside (the bar) was done in the steam-tug Fearless. I did not dare attempt working on the Bar in an open boat at this season of the year. The weather was much against me in executing this part of the work. I only succeeded in getting outside once in the tug, but of that I made as good use as possible. I remained out longer that the pilot desired, and in consequence, it was just the stage of extreme low water, spring tides, when we crossed the bar to return. We then found 11 feet on the bar. The tug was drawing 9? feet! There were 3- or 4-foot breakers continually rolling over the bar and we thumped the whole way over. After that made repeated attempts to go out again, but the seas were so heavy that the pilots refused to incur the risk.*

Other reports in those early years about the depth at the Coos Bay bar suggested a high-tide marker of fifteen feet with a possible depth of nineteen feet.

C. R. Smith arrived in Coos Bay during a period of important change for the port. In 1904, Coos County coal production had reached its peak. Coal shipments from the Coos County mines had increased gradually between 1880 and 1900, but year-to-year production had fluctuated widely. During those years, the local mines were competing with those of Puget Sound, where vessels drawing eighteen feet could call; Coos Bay could not ensure reliable passage to ships of that draft.

In September 1899, the US Chief of Engineers ordered a preliminary survey of the upper Coos Bay harbor. Captain Thomas W. Symons of the Corps of Engineers found several shoals and that impeded traffic except at high tide. Some shoals presented serious problems. To remove these

Crossing the Coos Bay Bar in the early 1900s (Coos County Historical Museum)

The 150-foot, three-masted schooner Advent*, built at the Simpson shipyard in North Bend in 1901, wrecked on the Coos Bay Bar on February 18, 1913. All men on board were rescued (Coos County Historical Museum)*

The C. A. Smith *under construction in 1917 at Coos Bay (University of Oregon)*

navigational barriers, Captain Symons recommended building a combination dredge and snag puller for use in the Coos River and the sloughs, as well as the Bay.

Congress, through the River and Harbor Act of March 2, 1907, approved a formal survey of the Coos Bay harbor; the study's final recommendation was to create an eighteen-foot channel from the jetty to the town of Marshfield, thirteen miles inside the Bay.

From 1908 to 1910, the dredge *Oregon* successfully completed the task of deepening and widening the channel progressing from the entrance to the C. A. Smith mill at milepost 16, where the Georgia-Pacific sawmill is currently located. Later the Corps added a pump dredge and piped the material for fills. Gradually the dredges deepened the channel all the way up the estuary to the mouth of Isthmus Slough. Ultimately, especially for the lower Bay, the engineers recommended a hopper dredge. The government subsequently built the *Colonel P. S. Michie* in Seattle and the dredge was launched on August 16, 1911.

After the channel was adequately deepened, in 1917 C. A. Smith had the Kruse and Banks Shipbuilding Company build a large ocean freighter (275 feet in length and 1,878 tons)—which he named after himself—to carry lumber to markets in San Francisco and other major ports of call. Unfortunately the ship sank after striking the north jetty of Coos Bay on December 16, 1923, while attempting to cross the bar. Of the fourteen-member crew, ten were saved.

Entrance to Coos Bay (ca. 1930) (Coos County Historical Museum)

Dredging, deepening, and other improvements to Coos Bay would serve C. R. Smith's Menasha shipping needs. But of equal—and perhaps greater—concern was access to tracts of timber. Here Menasha's interests became tangled with the complex laws governing public and private land ownership in the West.

The rock used to construct the jetties at the mouth of Coos Bay came from quarries along the Coos River. The large boulders were transported by barge to a staging area on the north and south sides of the channel, away from the force of the ever-present waves. From there, the rock was off-loaded onto rail cars for a short trip out the jetty where the stones were placed by a crane (Coos County Historical Museum)

The development of western land has been from the beginning of our nation, and still remains, an extremely divisive political issue. Thomas Jefferson in 1776 wrote: "The people who will migrate to the westward will be a people little able to pay taxes. . . by selling the lands to them you will disgust them and cause an avulsion (forcible separation) of them from the common union. . . . I am at the same time clear that they should be appropriated in small quantities."

"Small quantities" would not serve the needs of the timber industry. Sixty-five years after Jefferson's words, Congress determined that the national interest could best be served by transferring public domain lands by gift and sale to states, individuals, and companies, which would develop the western lands and its resources. Congress also believed that the only way to bind the vast nation together was to encourage the building of railroads and strategic military wagon roads. A variety of laws following the Civil War were designed to accomplish these goals.

As early as the mid-1850s the federal government began passing laws that would grant access to the public domain in order to give ownership to settlers and developers of the western frontier. Many of these land grants would lead to some of the most contentious and protracted legal battles over land this nation has ever seen.

John Lennan (left) and Del Lennan (right) hand-hewing a keel at Simpson's shipyard, North Bend, 1900 (Coos County Historical Museum)

During its period of western expansion, C. R. Smith's Menasha Wooden Ware Company had to pick its way through a maze of overlapping and sometimes conflicting land laws, the most important of which were:

A crew at Asa Simpson's North Bend shipyard. The white horse was a "horse of power" around the yard to haul large timbers and raise the timbers with block and tackle (Coos Art Museum, Victor C. West Collection)

Simpson shipyard at North Bend (ca. 1900) (Coos County Historical Museum)

The Kruse and Banks shipyard on Coos Bay (ca. 1920s), located at the site currently occupied by the Coquille Indian tribe Mill Casino, previously the large Weyerhaeuser sawmill site (Courtesy of Mary Granger)

COOS BAY SHIPBUILDING

Early Coos Bay lumberman Asa Simpson was also a master shipbuilder who capitalized on the magnificent timber available near his mill on Coos Bay. The tall, straight conifer trees provided excellent material for spars while the abundant old growth Douglas-fir was sawed for ship timbers and planking.

Captain Simpson's first ship went down the launching ways in 1857 (although his first lumber went out in 1856). The first major vessel built in 1859 at the Simpson shipyard on Coos Bay was the brig "Arago". It was 103 feet long and was the first brigantine to be built in the United States.

Shipbuilding was a necessity if the area was to succeed as a major lumber port. In the early days, the ships that called on Coos Bay were found to be of insufficient size to carry the long timbers originating in the mills. One solution was to create temporary holes in the stern and bow of the ships so the long timbers could be loaded. Asa Simpson was quickest to seize upon the opportunity to build ships more compatible with the lumber being produced on the Bay, and his shipyard business flourished. By 1902, fifty-eight ships had been built at Simpson's mill and shipyard at North Bend.

The hulls of these cargo ships were generally built of Douglas-fir and white cedar and rigged as two-masted schooners or brigantines. Soon the increased lumber production caused a demand for larger ships and four- and five-mast schooners were constructed.

Simpson's ships dominated the coastline from Coos Bay to San Francisco for decades. But he was not the only major shipbuilder in Coos Bay. The Kruse and Banks Shipbuilding Company, started in 1904, provided competition, constructing some of the largest ships ever made in Oregon.

John Kruse (Courtesy of Mary Granger)

Robert Banks (Courtesy of Mary Granger)

Kruse and Banks Shipyard (ca. 1920) (Coos County Historical Museum)

The Donation Land Claim Act of 1850, which stated: "There shall be and here by granted to every white settler or occupant of the public lands, American Half-breed Indians included, above the age of eighteen years, being a citizen of the United States. . . and who shall have resided upon and cultivated the same for four consecutive years, the quantity of one half section, or three hundred and twenty acres of land," and further: "That to all white male citizens of the United States . . . emigrating to and settling in the Territory [Oregon] between 1850 and 1853 . . . [shall be] granted the quantity of one quarter section or 160 acres," and further: "And upon the decease of the claimant, all the rights. . . shall descend to his heirs at law." This Act played a role in the land disputes for the cities of Marshfield and Empire City on Coos Bay, and ultimately had an impact on Menasha.

The Homestead Act of 1862, which authorized unrestricted settlement on public lands to all settlers, requiring only residence, cultivation, and some improvement of a tract of 160 acres. After living on the land for six months, the settler could buy the homestead for $1.25, and after living on the land for five years could apply for and receive a patent or title to the 160 acres for $15. This law in some ways opposed the Oregon and California Rail Road and Coos Bay Military Wagon Road land grants, creating a conflict that would ultimately embroil Menasha in a seven-year-long legal morass with the federal government.

The Timber and Stone Act of 1878 authorized the negotiated sale (at $2.50 per acre) of public lands (160 acres) especially valuable for either timber or stone and otherwise unfit for cultivation. The Act was repealed in 1955. The bargain land price was a bonanza to Douglas County lumbermen (Douglas County neighbors Coos County to the east). As Oregon historian Dow Beckham put it, "For the price of a good log, they could buy an acre of virgin forests! Although the law required taking an oath that the land was not taken in 'speculation,' tens of thousands of acres in western Oregon were illegally filed upon by persons who had been hired for $50, or perhaps a glass of beer, for making the land entry."

The Oregon and California Railroad Grants, which created what are now commonly called O&C lands, were made shortly after the Civil War. Congress provided subsidies in the form of vast land grants from the public domain to various states for the purpose of aiding the construction of railroads and to encourage westward expansion. For instance, Congress granted specific lands to the State of Oregon in 1866 for the construction of a railroad from Portland south to the California border near Ashland. The land grants allowed the state to select a private company to construct the railroad and to receive the land as a subsidy to offset the construction costs. The company could then sell the land to recover the "up-front" expenses of the construction. The amount of land involved was staggering. The O&C railroad land grant included all odd-numbered sections (640 acres each) of public domain land for twenty miles on each side of the proposed railroad line. If lands within the corridor were already claimed by homesteaders, the company's corridor would be extended to thirty miles on each side of the rail line. The total amount of the land granted in Oregon under O&C was 4,220,000 acres—an area larger than the state of Connecticut. This program of developing the west through railroad land grants would end up in the Supreme Court of the United States and before the US Congress, which would ultimately unravel the fifty-year quagmire surrounding the ownership of these lands.

The Ed Bunnell residence, Coos County, Oregon, 1889 (Courtesy of Gordon Ross)

Old-growth Douglas-fir forest (University of Washington)

With the continued improvement of the Wagon Road, more elegant forms of transportation were used for the two-and-a-half-day trip across the Coast Range (Douglas County Historical Museum)

Road construction in the late 1800s was mostly done by animal and human power. Men and teams with scrapers and a wagon shown here grading the Coos Bay Military Wagon Road near Brewster Canyon (Douglas County Historical Museum)

Stagecoach stations with overnight accommodations cropped up along the road for the weary travelers. The Fairview Stage Station and post office on the Wagon Road in 1890. Note in the right corner of the photo a water trough for horses hewn from a log (Douglas County Historical Museum)

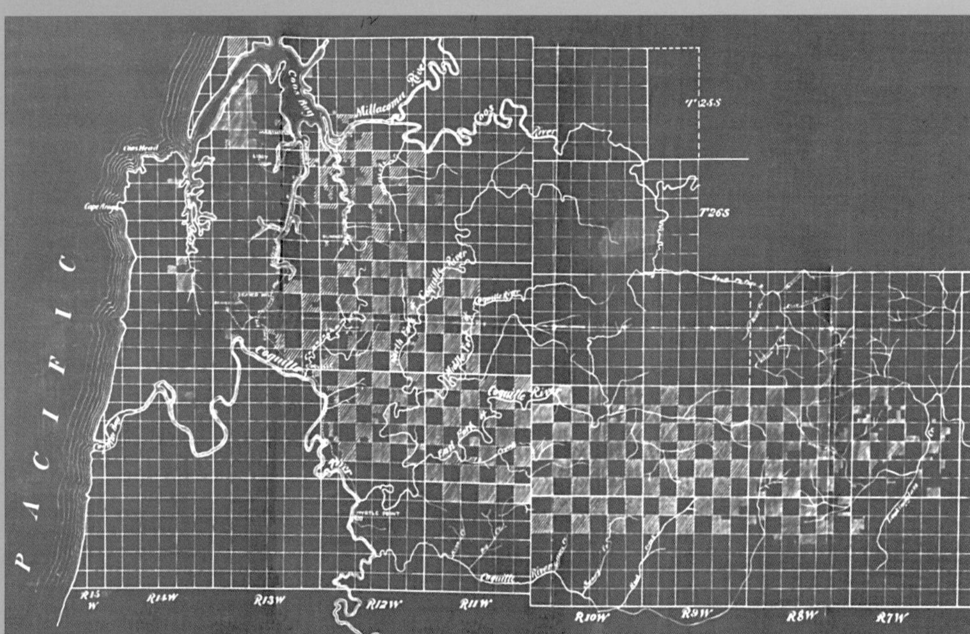

Map of the Coos Bay Wagon Road lands and those of Henry Luse. Both tracts would transfer from the Oregon Southern Improvement Company to the Southern Oregon Company, and finally to Menasha in 1908. The shaded area totaled 115,000 acres

First auto to travel the Coos Bay Wagon Road, 1908 (Douglas County Historical Museum)

Military Wagon Road Grants began in Oregon in 1864 when Congress, following the Civil War, decided it was important for the nation to have a better road system between forts and important shipping points. It granted lands to the state for later transfer to a private road company for the purpose of constructing four military roads in Oregon. The last of these, the Coos Bay Military Wagon Road, was created by Congress on March 3, 1869, to connect the coastal shipping port with its customs office to the more populated interior valleys of the state. As with railroad grants, road companies were given enormous tracts of land with massive stands of timber in lieu of cash from federal and state governments to cover the cost of road construction. The road company could select odd-numbered sections of land within three miles on each side of the surveyed roadway from the public domain for every mile of roadway it built. The Coos Bay Wagon Road Company was organized in 1870 to build the road from what was then known as Coos City (some four miles south of Marshfield) eastward to the town of Roseburg, Oregon, and received the land grant from the State of Oregon.

The road officially opened in 1873 and qualified the company owners to a land subsidy of 105,120 timber-rich acres.

Available timberland on this vast scale was just what Menasha and other timber companies needed. Competition for access to the railroad and military road land grants started almost immediately.

The construction of the Wagon Road proved far more difficult than the Coos Bay Wagon Road Company anticipated. Not only did the general topography present challenges, but the rock outcrops and large old-growth timber stumps had to be removed. Animal power was again the only real "muscle" around.

The Coos Bay Wagon Road Company found it could not adequately maintain the roadway and in 1875 sold some of its rights to a group of San Francisco financiers that included Leland Stanford. On December 23, 1883, a group of investors and developers, the Oregon Southern Improvement Company (OSIC), gained control of the road. Four years later the property was sold at a sheriff's sale to the Southern Oregon Company—an investment group that included a number of men who had formed the OSIC.

The combination of federal land grants, homesteading laws, and business dealings led to the transfer of some thirty million acres of public domain lands in Oregon to individuals, companies, or the state itself. That totals just less than half the land in Oregon. Jerry A. O'Callaghan in his treatise "The Disposition of the Public Domain in Oregon" showed how the Public Domain lands in Oregon were disposed:

Homestead	11,097,982 acres
Land Sales	6,455,551
Transition Grant to States (RR and schools)	4,329,445
Donation Land Claims	2,614,082
Wagon Road Grants	2,490,890
Railroad Grants	1,588,532
Miscellaneous	992,921
Total	29,569,921 acres

Coos City, the beginning of the Wagon Road on Isthmus Slough (Douglas County Historical Museum)

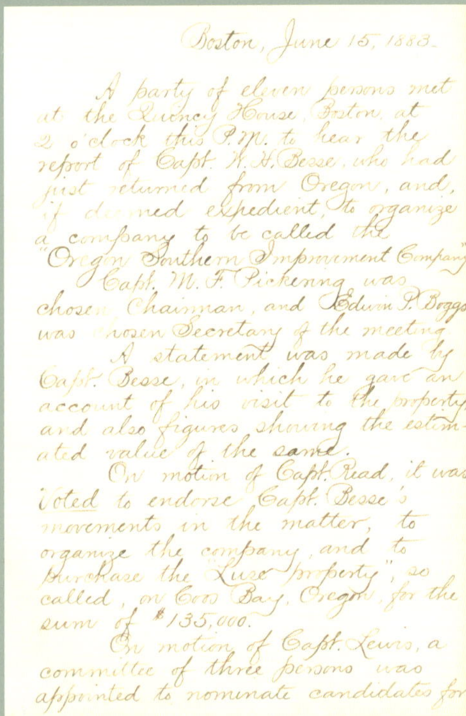

Boston, June 15, 1883.

A party of eleven persons met at the Quincy House, Boston, at 2 o'clock this P.M. to hear the report of Capt. W. H. Besse, who had just returned from Oregon, and, if deemed expedient, to organize a company to be called the "Oregon Southern Improvement Company"

Capt. M. F. Pickering was chosen Chairman, and Edwin P. Boggs was chosen Secretary of the meeting.

A statement was made by Capt. Besse, in which he gave an account of his visit to the property, and also figures showing the estimated value of the same.

On motion of Capt. Read, it was Voted to endorse Capt. Besse's movements in the matter, to organize the company, and to purchase the "Luse property", so called, on Coos Bay, Oregon, for the sum of $135,000.

On motion of Capt. Lewis, a committee of three persons was appointed to nominate candidates for

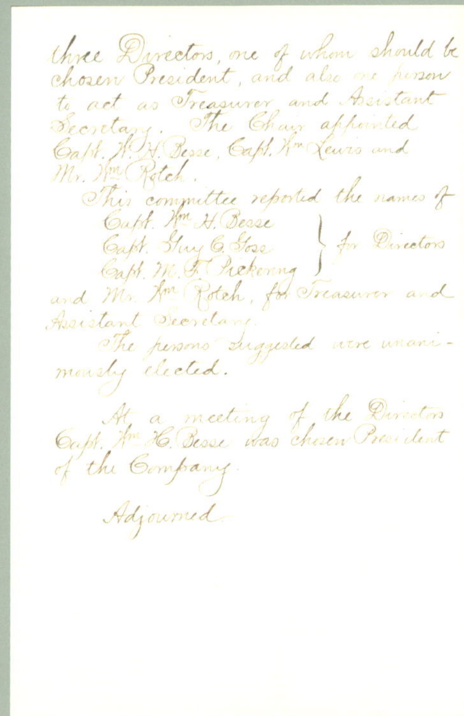

three Directors, one of whom should be chosen President, and also one person to act as Treasurer and Assistant Secretary. The Chair appointed Capt. W. H. Besse, Capt. Wm Lewis and Mr. Wm Rotch.

This committee reported the names of

Capt. Wm H. Besse }
Capt. Guy C. Goss } for Directors
Capt. M. F. Pickering }

and Mr. Wm Rotch, for Treasurer and Assistant Secretary.

The persons suggested were unanimously elected.

At a meeting of the Directors Capt. Wm H. Besse was chosen President of the Company.

Adjourned.

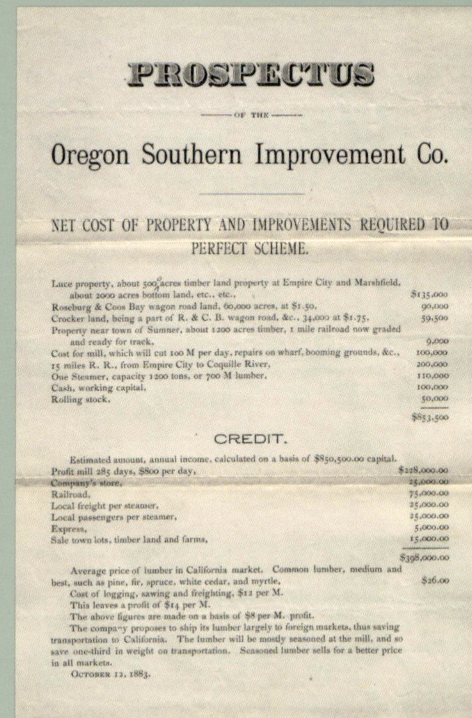

PROSPECTUS

OF THE

Oregon Southern Improvement Co.

NET COST OF PROPERTY AND IMPROVEMENTS REQUIRED TO PERFECT SCHEME.

Luce property, about 5000 acres timber land property at Empire City and Marshfield, about 2000 acres bottom land, etc., etc.,	$135,000
Roseburg & Coos Bay wagon road land, 60,000 acres, at $1.50,	90,000
Crocker land, being a part of R. & C. B. wagon road, &c., 34,000 at $1.75,	59,500
Property near town of Sumner, about 1200 acres timber, 1 mile railroad now graded and ready for track,	9,000
Cost for mill, which will cut 100 M per day, repairs on wharf, booming grounds, &c.,	100,000
15 miles R. R., from Empire City to Coquille River,	200,000
One Steamer, capacity 1200 tons, or 700 M lumber,	110,000
Cash, working capital,	100,000
Rolling stock,	50,000
	$853,500

CREDIT.

Estimated amount, annual income, calculated on a basis of $850,500.00 capital.

Profit mill 285 days, $800 per day,	$228,000.00
Company's store,	25,000.00
Railroad,	75,000.00
Local freight per steamer,	25,000.00
Local passengers per steamer,	25,000.00
Express,	5,000.00
Sale town lots, timber land and farms,	15,000.00
	$398,000.00

Average price of lumber in California market. Common lumber, medium and best, such as pine, fir, spruce, white cedar, and myrtle, $26.00

Cost of logging, sawing and freighting, $12 per M.

This leaves a profit of $14 per M.

The above figures are made on a basis of $8 per M. profit.

The company proposes to ship its lumber largely to foreign markets, thus saving transportation to California. The lumber will be mostly seasoned at the mill, and so save one-third in weight on transportation. Seasoned lumber sells for a better price in all markets.

October 12, 1883.

Henry Luse wharf and store, purchased in 1883 by the Oregon Southern Improvement Company (Oregon Historical Society 684-A)

The relevance of these land grants to Menasha is simple. The transfer of public lands to private hands built many fortunes in early Oregon history. When C. R. acquired all of the stock of the Southern Oregon Company by 1911, the disposition of the Coos Bay Wagon Road lands would provide much of the initial western timberland ownership for his company. But establishing that control would prove only temporary—and would land C. R. Smith smack in the middle of a series of early twentieth-century debates and lawsuits surrounding the disposition of these western public domain lands.

Tracking the land is vital in understanding the history of Menasha. That history begins with the tangled interests of a number of people and companies who came to the Coos Bay area to cut timber in the last half of the nineteenth century.

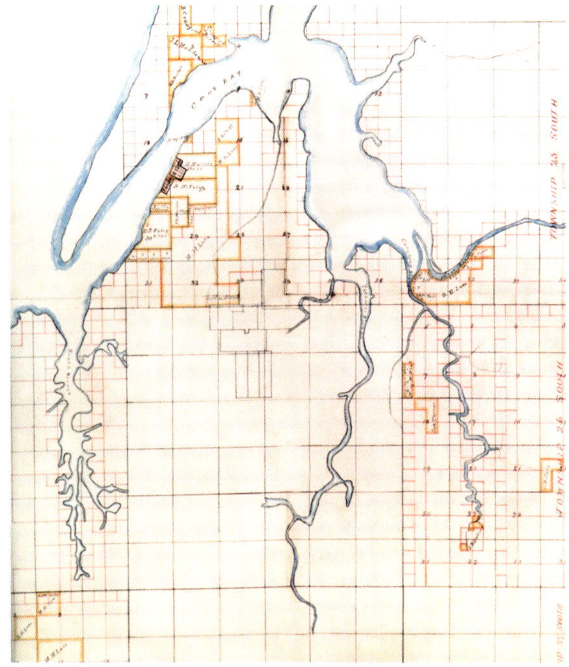

The Henry Luse property (blocked out in orange), acquired by the Oregon Southern Improvement Company in 1883. The town site of Empire City is also shown.

When C. R. Smith began buying timberland in the area in 1905, he started dealing with businessmen who had already been involved in the industry in Coos Bay for a generation. We have met some of the biggest: Asa Meade Simpson, his son Louis, and Henry Heaton Luse, major owners of timberlands, mills, and shipyards in the area. But they were not the only major players.

In early 1883, Captain William Besse of New Bedford, Massachusetts, who was on the Pacific Coast in connection with his shipping interests, learned of the properties of Henry Luse and the Coos Bay Wagon Road Company. Likely inspired by the hyper-speculative atmosphere of the times, he returned to New England and encouraged a group of business associates to acquire the Wagon Road land. On June 15, 1883, they moved to organize the Oregon Southern Improvement Company to do so.

The first records of the OSIC can be traced to the minutes of their original meeting at the "Quincy House" in Boston, Massachusetts, on June 15, 1883. A group of eleven distinguished men from Boston and New Bedford had gathered at the house to hear Captain Besse's account of his recent visit to Coos Bay, Oregon. They quickly determined that the value of the lands and timber offered them a chance to make a profit, and began the process of raising investment money.

The founding subscribers to the OSIC were:

- William H. Besse, President
- G. C. Goss
- M. F. Pickering
- A. G. Page
- J. L. H. Cobb
- B. S. H. Cook
- C. W. Lewis
- George S. Homer
- D. A. Besse
- E. W. Seabury
- William J. Rotch

Southern Oregon Company railroad at Sumner, Oregon, 1888 (Coos County Historical Museum)

The wooden steamer AL-KI *docked at the OSIC wharf in Empire City (ca. 1887) (Coos Art Museum, Victor C. West Collection)*

Inside The Big Mill (Coos County Historical Museum)

The OSIC raised $235,000 in cash and dispersed $100,000 toward the purchase from Luse of 5,000 acres of timberland and 2,000 acres of bottomland along with platted lots in Empire City, two railroad cars, and a sawmill on Coos Bay. The final deed was recorded in Coquille on November 22, 1883.

Four months later, the OSIC board authorized the construction of the steamer *AL-KI* to be built in Bath, Maine, for use in transporting goods from the East coast to their new business ventures in Coos Bay. She arrived in the Bay on November 21, 1884, with a cargo of supplies for the company store. As it turned out, the *AL-KI* was too big for the draft at the bar at Coos Bay for regular shipments. The ship wrecked at Point Augusta, Alaska, on November 3, 1917.

At a special meeting of the OSIC Board of Directors on February 1, 1884, the Board voted to ratify the purchase of the franchises and property of the Oregon Southern Railway and Transportation Company for $5,000 cash, giving the group rail access for the transportation of logs from their timberlands in the Sumner, Oregon, area to Empire City for milling and shipping.

During that same meeting, the board also voted to acquire 94,000 acres from the Coos Bay Wagon Road Company through the use of stocks and bonds totaling $800,000 of the company.

The OSIC Board of Directors held its first meeting at the new company office in Empire City, Oregon, on July 24, 1884, at which time Elijah Smith (again, no relation to Menasha's founder, Elisha Smith) was elected president, William P. Granger vice president, George H. Miner (no relation to Willis Miner who would become Menasha's president forty years later) secretary, and Prosper W. Smith (no Menasha relations, either) as treasurer.

With their acquisition of the Luse property and the 94,000 acres of Wagon Road lands, a new sawmill was needed. At their Board meeting on June 14, 1884, in Boston, the directors authorized William P. Granger to take charge of their properties and to "superintend the erection of a sawmill in Empire City." The Big Mill, as it came to be called, would become an important piece of the area's history.

Later that month Granger wrote the OSIC treasurer requesting payment for drafts he had written for the purchase of machinery for the mill: engines and boilers, $24,200; circular saws, $9,604. He also reported contracting for the harvesting of one million board feet of logs from company lands as follows:

Log lengths	Logging costs
16–36 feet	$4.50/MBF
37–50 feet	$5.00/MBF
51–60 feet	$6.00/MBF
61–70 feet	$7.00/MBF
71–80 feet	$8.00/MBF
81–90 feet	$9.00/MBF
91–100 feet	$10.00/MBF
101–110 feet	$11.00/MBF
111–120 feet	$12.00/MBF
121–125 feet	$12.50/MBF

The Oregon Southern Improvement Company mill at Empire City (ca. 1885) (Coos Art Museum, Victor C. West Collection)

Largest log ever take from the Davis Slough area: 182 feet long, 7.5 feet across at the butt, and 5.5 feet across at the top (1910). Menasha owns more than 10,000 acres in the Davis Slough drainage of Coos County (Coos County Historical Society)

The Cape Arago Lumber Company (the former OSIC Big Mill) lumber storage area in the 1950s. The ship at the Empire fuel depot dock is a converted Olsen steam ship used as a troop carrier during World War II (University of Oregon)

A 1902 report found in the Menasha office in North Bend includes the following description of the OSIC Big Mill:

> *The sawmill had two 60-inch circular saws capable of sawing a log 60 inches in diameter and 40 feet long. The triple mill can saw up to 10 feet in diameter and 130 feet in length! The double mill has a steam feed and the triple has a wire rope feed, driven by independent engines. Also a Wickes gang saw, 26 inch stroke, 48 inch gate, two improved gang edgers, 3 planers, 1 molding machine, 5 cut-off saws, a shingle mill, and a barley mill. The output of the mill was estimated, with both sides running, at about 150,000 board feet lumber scale per 10-hour day.*
>
> *The mill is supplied with automatic transfers facilitating the handling of lumber and with sawdust conveyors throughout the building. It also has an electric light plant supplying light to the mill, store, and office buildings.*

The original cost of the Big Mill was about $150,000, but more improvements were added in later years. It was undoubtedly one of the era's largest West Coast mills. The building was constructed of heavy sixteen-by-sixteen-inch timbers in lengths ranging from fourteen to fifty-two feet. They were probably cut in the old Luse sawmill next door. Pile driving and framing for the big mill started in July 1884. Granger purchased one seventy-two-foot long timber from the E. B. Dean and Company sawmill at Marshfield and paid $120/MBF. According to Dave Miller Sr., who was manager and final owner of this mill (renamed the Cape Arago Sawmill) before it was demolished in the 1980s, the mill was equipped with a ten-foot head rig and the big underpinnings that supported the big saw were twenty-by-twenty-inch hewn timbers.

The mill ran on steam from 1883 until 1975 when it was converted to air and electricity by the Cape Arago Lumber Company. A four-inch wooden pipe that ran from Empire lakes to the mill supplied water for the steam plant. Dave Miller recalls the engineering of the building was remarkable as there were few nails and the beams were secured together with wooden pegs.

THE RISE AND FALL OF EMPIRE CITY

When the Oregon territorial legislature created Coos County on January 19, 1854, the only platted town and center of commercial activity in the region was a place few of today's Oregonians have ever heard of: Empire City. It was the first Coos County seat, home of the first county courthouse and an official state port of entry. The area was rich in timber, fish and coal, and there was talk of a major roadway and railroad reaching into the region. In 1873, with the illusive promise of mounting commerce as a consequence of the construction of the Coos Bay Wagon Road, Congress named Empire City a port of entry for the United States and opened a customs house in the town. In 1884, the Oregon Southern Improvement Company (OSIC) built its Big Mill and started shipping lumber. The future looked bright.

But the city's development never achieved its potential. One local newspaper reporter in 1918 blamed it on the Southern Oregon Company (SOC), which took over OSIC's Empire City properties after the latter group went broke: "The [Southern Oregon Company] killed Empire," the reporter wrote. "In an attempt to hog everything, it refused to sell town lots at prices that would let other industries than its own to live. It maintained the same attitude toward would-be buyers of residence lots."

Eventually Empire City was eclipsed by the development, farther up the Bay, of Marshfield (which would become officially known as Coos Bay in 1944) and North Bend (which was originally called Yarrow until Louis B. Simpson had the name officially changed in 1902). In 1896, voters moved the county seat from Empire City to Coquille, and the future of Empire City was forever changed.

However, moving the courthouse would not occur without a legal battle. It took a meeting of the Coos County Commission to find a funding method that would negate a lawsuit brought by the OSIC against the move (it was in the company's interest that Empire City remain the county seat). The commissioners decided to build the new courthouse in Coquille with traditional funding methods, according to Gordon Ross, current Coos County commissioner (Note: Gordon Ross's grandfather was one of the elected officials on the Commission in 1897!).

Empire City would not die without a fight. In January 1927, W. G. Robertson, along with Louis Simpson and others, announced they had filed incorporation papers for the Empire Development Company. Rumors circulated that another mill and a town site were planned. By April 1927, the company had purchased 1,600 acres of local timberland as well as all of the residential town lots in Empire City originally owned by the SOC. By June of that year, they had purchased another 1,000 acres northeast of the town. In December of 1928, the Empire Development Company directed Louis Simpson to rebuild the old OSIC Big Mill; that portion of the business was called the Empire Western Lumber Company and was later reorganized into the Simpson Lumber Company. Unfortunately, Simpson went into receivership during the Depression of 1929 and the Southern Oregon Company (later to become part of Menasha) repossessed the mill.

Menasha records indeed show that the SOC (then a subsidiary of Menasha Wooden Ware Company) sold the mill in Empire City to the Empire Development Company, but Carlton Smith from Menasha's corporate office in Wisconsin must have seen the writing on the wall. In a 1928 internal company memo, he noted as much, using language like, "Inasmuch as the Empire Lumber Company situation is somewhat precarious at this time and there is a question as to whether they can complete their contract for purchasing this mill and also inasmuch as the Empire Development Company's position is more or less unstable. . . ."

Menasha secured a quitclaim deed back from the Empire Development Company for the mill site and once again the future of Empire City was thrown into doubt.

After C. Johnson, James A. Lyons, and Robert R. Dutton acquired the rights to the Empire Lumber Company, they leased the OSIC Big Mill in Empire City and reopened it between 1933 and 1938. According to Jim Lyons, grandson of James A. Lyons, the Empire Lumber Company fell on hard times; as the family story is told, the company was one step ahead of a sheriff's sale for the equipment in the mill to satisfy a loan issued by a Portland financier. Lyons and partners had moved the equipment in the mill over to another sawmill building called Mill B just south of the town of North Bend. The Big Mill in Empire went back to Menasha, again, on August 8, 1938.

The Cape Arago Lumber Company, which purchased the mill in October 1940, ran it successfully for many years. Throughout these roller-coaster times, there was little that Empire City could cling, except for this mill facility.

Today, shorn of its "City," the renamed Empire exists as an outlying part of the city of Coos Bay-North Bend, its heyday more than a century in the past.

Carlton Smith, grandson of Elisha D. Smith

Henry Luse's steam powered tug, Alpha, being built at Empire City in 1868. (Coos Art Museum, Victor C. West Collection)

The Oregon Southern Improvement Company sawmill in Empire during the summer of 1884 (Coos Art Museum, Victor C. West Collection)

Empire city (ca. 1890)

Empire City (ca. 1890). The big Oregon Southern Improvement Company sawmill building can be seen at the far left. The original Henry Luse sawmill had been dismantled by this time

Looking south across Empire City (ca. 1890) along the waterfront where the Henry Heaton Luse and the Oregon Southern Improvement Company owned five-to-seven miles of upland Bay frontage and the tidelands abutting them

1930 aerial photo of Empire City, Oregon . Note the old Oregon Southern Improvement Company mill in upper left (Coos Art Museum, Victor C. West Collection)

Logging camp bunkhouse in Clallam County, Washington, 1905 (Cowlitz County Historical Museum)

Harvesting costs for logs cut on the OSIC lands near Empire City ranged from $4.50 to $5.00/MBF depending upon length. Captain Knowles was in charge of the logging camp for the company. Granger also purchased logs from the local loggers that were delivered directly to the company booms at $4.50 to $5.00/MBF. By November 1884, the mill had purchased 4,537 MBF of logs for $22,451, or $4.95/MBF. (Granger's descriptions of the Big Mill construction and operation are in the files at the office of Menasha Forest Products Corporation in North Bend, Oregon.)

Looking back, the Big Mill had a bit of a roller-coaster ownership after its construction.

Luse Mill	1856
OSIC Big Mill construction	1885
Sale to SOC	1887
George Loggie modernization*	1888
Running on-and-off	1885-1892
Plant idle	1893-1927
Empire Development Co.**	1927
Sale to Western Empire Lumber Co.***	1928
Sale to Simpson Lumber Co.	1928
Mill restarts	1929
Menasha takes it back	1933
James Lyons leases mill for one year	1935
Sale to Cape Arago Lumber Co.	1941-1980
Mill dismantled	1980

* George W. Loggie, a prominent sawmill man from Washington and a director of the subsequent Southern Oregon Company, further modernized the big mill.

** One of the owners of The Empire Development Company was Louis J. Simpson, who was assigned the job of rebuilding the old OSIC Big Mill. He purchased the machinery from the Ferry-Baker Lumber Company in Everett, Washington, and moved it to the Empire City site.

***Louis J. Simpson owned Western Empire Lumber Company.

Dave Miller confirms that Irwin and Lyons moved their sawmill equipment out of the building in 1939 as the sheriff was headed to the mill to padlock the door for the creditors. Jim Lyons, grandson of the partner in Irwin and (James A.) Lyons acknowledged that his grandfather and Irwin had borrowed $150,000 for the mill equipment and essentially went broke. They took the machinery to what was then known as Mill B in North Bend and installed it there.

The OSIC had the land, the timber, and the mill to become a major player in the area's economic future. But within a very short time, the OSIC would be gone, replaced by a new organization that would forge a direct link with C. R. Smith's Menasha Wooden Ware Company.

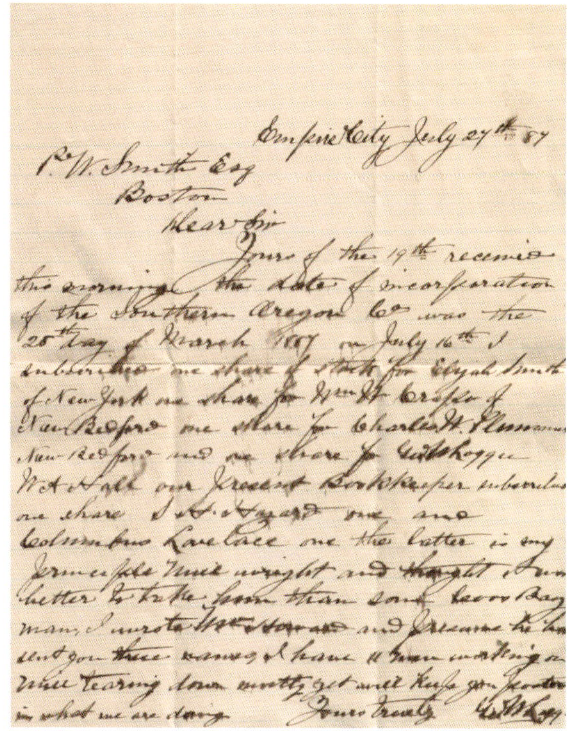

Empire City, July 27th (18)87

P.W. Smith Esq

Boston

Dear Sir

Yours of the 19th received this morning the date of incorporation of the Southern Oregon Co was the 25th day of March, 1887 on July 16th I subscribed one share of stock for Elijah Smith of New York one share for Wm W Crapo of New Bedford one share for Charles W. Plummer New Bedford and one share for Geo W Loggie W.A. Hall our present bookkeeper subscribed one share S.H. Hazard one and Columbus Lovelace one the latter is my principle mill wright and thought it was better to take him than some Coos Bay man. I wrote Wm Howard and presume he has sent you these names. I have 11 men working on mill tearing down mostly yet will keep you posted in what we are doing

Yours truely(sic) Geo W Loggie

Transcription of original note from George Loggie (original note above)

EARLY LOGGING IN COOS COUNTY

Logging in the early days around Coos Bay was quite a challenge. Aside from the steep topography, there were very few good roads reaching into the backcountry and steam power was a novelty in a logging operation. Oxen teams were the primary method used to drag logs from the stump (often over long distances) to a watercourse or local mill. Trees were large and only the most valuable ones would make it out of the woods and into the mills. Production was incredibly slow by today's standards. Timber falling was done with a double-bitted axe and a "misery whip," or two-man crosscut saw.

To eliminate most of the problems associated with stump rot in the "butt swell" of old trees, to save the extra effort required to saw through the wide base of the unusable wood, and to produce logs that were fairly round, timber fallers would cut small notches in a tree, insert "spring boards" into the notches and climb up high enough to get above the irregular butt swell to make their cut

As animal power was the only means of getting these big trees to the mill, only the best quality logs were taken.

It is hard to imagine the force with which these big trees crashed to the ground. Despite the most diligent efforts of the professional fallers, breakage was inevitable. Trees might land on recently cut stumps, fall across logs (old or new), or hit a small boulder hidden by the brush, any one of which could shatter a section of the tree.

Once the big old-growth trees were on the ground, they had to be cut into smaller pieces that matched to sawmill capacity and the strength of the oxen team that had to pull them out of the woods. Bucking the logs was a laborious and dangerous process that was done by hand until the advent of gas-powered bucking saws. Once the trees were cut into logs, they had to be moved from the stump to an intermediate water or rail siding destination before making their final journey to the sawmill.

Logging camps in those early days were pretty basic. Men lived and ate in relative tight quarters and given the damp climate of the Coast Range, one could not survive the trials of the bunkhouses with a sensitive nose!

Log bucker (Oregon Historical Society)

This red cedar tree is being felled twenty feet above the ground to eliminate the difficulty of sawing through the fluted butt and the problems that the irregularly shaped log end would create when being dragged to the water's edge by a team of oxen (University of Oregon)

Falling the big ones using a misery whip (crosscut saw that could range from eight to twelve feet in length), "spring boards" (wooden planks with steel "toes" for inserting into a notch in the tree where the fallers are standing), a double-bitted axe, and falling wedges (Douglas County Historical Museum)

Prime old-growth Douglas-fir logs being readied for skidding to the water's edge. Note the remaining trees left standing as loggers judged they were of no value. (University of Oregon)

An oxen team could pull large quantities of logs as long as the grade was flat or down hill (ca. 1890) (Coos County Historical Museum)

Here a bull team snakes logs down a ravine in Coos County. The man immediately behind the team is called the "bull puncher" or "bull whacker" and the others alongside the logs are called "skid greasers." They applied grease to the skid planks to help the logs slide easier. The greasers used whatever was available, from rendered fish oil or animal fat to tubs of rancid butter. If nothing else was available, simple water was applied to the skids (Coos County Historical Museum)

The R. E. Scranton narrow-gauge railroad was pulled by a steam engine in 1890 (Courtesy of Warren Dick)

The R. E. Scranton bull team in 1890 pulls a skid of logs to a narrow-gauge railroad in Coos County, Oregon (Courtesy of Warren Dick)

Once the logs were loaded on the Scranton narrow-gauge rail line, they could be hauled to a log-dump where they would be splashed into the river or the Bay and towed to a mill by a scow (Courtesy of Warren Dick)

The Coos Bay Military Wagon Road at Rocky Point between Roseburg and Coos Bay. The driver is Walter Laird (Douglas County Historical Society)

chapter three

The Wagon Road

The old Coos Bay Military Wagon Road lands included more than 100 square miles of Oregon timberland worth a fortune to whoever owned it. Battles over that land, including lawsuits, state political fights, a US Supreme Court ruling, and three changes of ownership between the mid-1880s and 1911, would have a single result: a portion of the Wagon Road lands would belong to C. R. Smith and Menasha—at least temporarily.

The Coos Bay Wagon Road had its inception on September 5, 1865, when the Coos County Court ordered:

> *. . . that the Road located from Huffman's Ferry on the North Fork of the Coquille river in Coos County, Oregon, running up the middle Fork of said River by the way of Enchanted Prairie, thence in an Easterly direction to the boundary line between Coos and Douglas counties, it is ordered that said road be let to the lowest bidder for the term of ten years, persons leasing said Road to make a good pack trail six feet wide in the clear, the grade of said road not to be more than ten feet in one hundred feet and if a good Wagon road is not made at the end of ten years from the line that the said pack trail is completed then the said lease to be forfeited, that three gates shall be placed on the Road, and it is ordered that sealed bids shall be received a the County Clerks Office for the leasing of said Road until Thursday the 28th day of December 1865 at 10 o'clock am and that notice is published in the* Oregon Statesman *four weeks.*

Apparently construction proved more difficult than anticipated. The initial contractor defaulted, even after additional funds were allocated. After two years without much progress, on July 2, 1967, the Coos County Court again ordered:

> *. . . that a road from the residence of Mrs. J. Huffman on the Coquille River. . . to the County line of Douglas County, be leased for the term of ten years to such party or parties as will open, improve and keep in repair the said Road upon the best of terms, following the location made by George Brown, Daniel T Hill and Thomas J. Rowley as filed in the county Court September 24, 1865. That the grade of said road shall not exceed four feet to the rod that there shall be no more than two toll gates on said road and that the bridges may be constructed of either wood or stone.*

The Federal Government became interested in the project and in 1869 made a grant of 99,819.35 acres of land to spur private construction of a road from Coos Bay to Roseburg. The federally sponsored route ran from Roseburg through Looking Glass and Ten Mile in Douglas

Once the road was built, the winter rains and the steep terrain wreaked havoc on the dirt roadbed (Douglas County Historical Museum)

Skid Road (Oregon Historical Society 7908)

Laird halfway house along the Wagon Road. The station burned on June 10, 1904 (Douglas County Historical Museum)

The mail was transported from Roseburg to Coos Bay over the Wagon Road as seen in this 1909 photo of Art Alexander and pack horses carrying the mail (Douglas County Historical Museum)

As more modern modes of transportation developed, the Wagon Road was improved even further as demonstrated by the first auto car to make the journey from Coos City to Roseburg in 1908 near Devil's Gap. Driver is Henry Harth, center is Eva Harth, and back is Al Dailey (the blacksmith) (Douglas County Historical Museum)

County, over the Coast Range to Sitkum on the East Fork of the Coquille River, and ended on Isthmus Slough at Coos Bay.

Although the road was completed in 1873, growing public discontent with road conditions and irregularities in the grant of land sales caused the Coos County Court to order on July 1, 1882:

> *That John Burnett was employed to investigate the report to this court before September term, whether or not the lands obtained by the Coos Bay Wagon Road Company from the United States for building a Wagon road from Coos City to Roseburg in this State, would not be made to comply with the conditions which they obtained said land.*

As noted in Chapter Two, the Oregon Southern Improvement Company owned the Wagon Road lands in the mid-1880s, and the company was making a substantial investment to become a major timber and lumber power in the Coos Bay area. But in the spring of 1887, as the OSIC was beginning to falter, a new player emerged: the Southern Oregon Company (SOC), formed by a group of Boston-area businessmen. The SOC's first official meeting (the records of which are stored today at the Menasha offices in North Bend) began with the following:

> *We the undersigned being all of the stockholders of the Southern Oregon Company, a corporation duly incorporated under the general laws of the State of Oregon, by articled of Incorporation dated on 25th day of March 1887, hereby unite in calling a first meeting of the stockholders of said corporation to be held at the office of Hazard and Wilson in the town of Empire City...on the 6th day of February 1888.*

Signatures to the subscription included:

Elijah Smith by proxy George W. Loggie
Prosper W. Smith " "
James W. Potter " "
William W. Crapo " "
William J. Rotch " "
Edward D. Mandell " "
William H. Besse " "
George S. Homer " "
Charles W. Plummer " "
George W. Loggie
S.H. Hazard
C.F. Lovelace
James Calderwood

Some of the names should sound familiar to the reader because several of the men founding the SOC were also among those who had started the Oregon Southern Improvement Company just a few years earlier. The OSIC, however, after spending hundreds of thousands of dollars on its mill, land, and shipping needs in Coos Bay, was going broke. From the records of the Coos County Clerk, it appears that OSIC gave a mortgage to all of their property to the Boston Safe

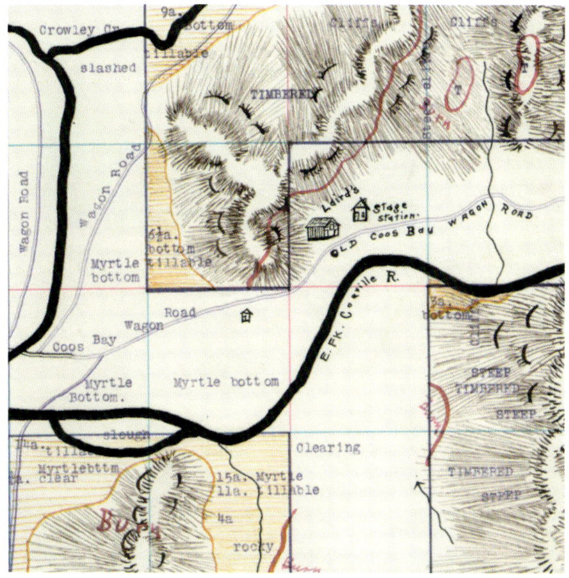

A 1912 hand-drawn map of some of the Coos Bay Wagon Road Grant lands in S11, T 28 S, R 10 W, Coos County, Oregon, as cruised by Lester Clarke in July of that year. He measured a total of 11,365,000 board feet of timber in the shaded areas

Today's foresters have far more extensive technology with which to evaluate tracts of forest land. This aerial photograph from 2002 gives a detailed "bird's-eye" view of the same area shown in the 1912 drawing

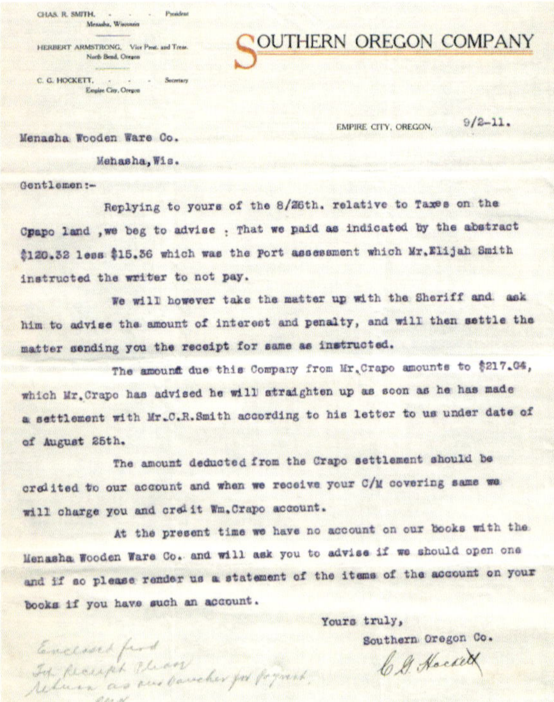

CHAS. R. SMITH, - - - President
Menasha, Wisconsin

HERBERT ARMSTRONG, Vice Prest. and Treas.
North Bend, Oregon

C. G. HOCKETT, - - - Secretary
Empire City, Oregon

SOUTHERN OREGON COMPANY

EMPIRE CITY, OREGON, 9/2-11.

Menasha Wooden Ware Co.
Menasha, Wis.

Gentlemen:-

Replying to yours of the 8/26th. relative to Taxes on the Crapo land, we beg to advise: That we paid as indicated by the abstract $120.32 less $15.36 which was the Port assessment which Mr. Elijah Smith instructed the writer to not pay.

We will however take the matter up with the Sheriff and ask him to advise the amount of interest and penalty, and will then settle the matter sending you the receipt for same as instructed.

The amount due this Company from Mr. Crapo amounts to $217.04, which Mr. Crapo has advised he will straighten up as soon as he has made a settlement with Mr. C.R. Smith according to his letter to us under date of of August 25th.

The amount deducted from the Crapo settlement should be credited to our account and when we receive your C/M covering same we will charge you and credit Wm. Crapo account.

At the present time we have no account on our books with the Menasha Wooden Ware Co. and will ask you to advise if we should open one and if so please render us a statement of the items of the account on your books if you have such an account.

Yours truly,
Southern Oregon Co.
C. G. Hackett

Southern Oregon Company letterhead, September 2, 1911 showing C. R. Smith as president, Herbert Armstrong as vice president and treasurer and C. G. Hackett as secretary

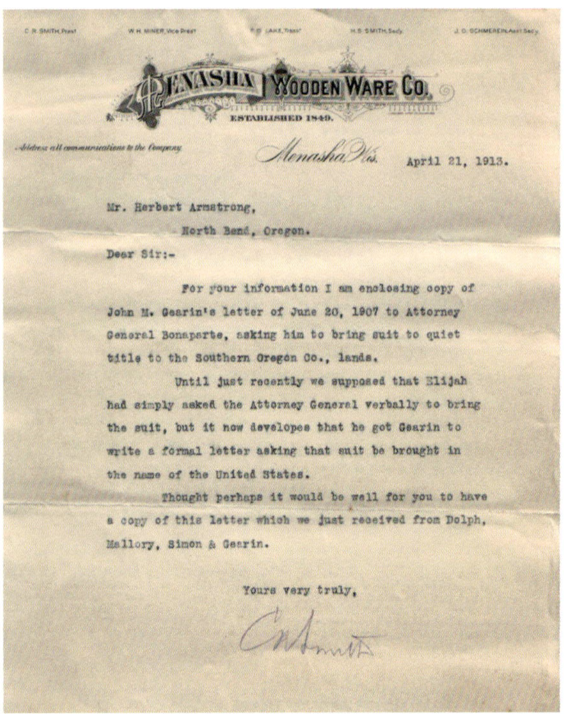

MENASHA WOODEN WARE CO.
ESTABLISHED 1849.

Address all communications to the Company

Menasha, Wis. April 21, 1913.

Mr. Herbert Armstrong,
North Bend, Oregon.

Dear Sir:-

For your information I am enclosing copy of John M. Gearin's letter of June 20, 1907 to Attorney General Bonaparte, asking him to bring suit to quiet title to the Southern Oregon Co., lands.

Until just recently we supposed that Elijah had simply asked the Attorney General verbally to bring the suit, but it now developes that he got Gearin to write a formal letter asking that suit be brought in the name of the United States.

Thought perhaps it would be well for you to have a copy of this letter which we just received from Dolph, Mallory, Simon & Gearin.

Yours very truly,
C R Smith

1913 letter written by C. R. Smith to Herbert Armstrong, Menasha manager in North Bend, discussing the ongoing lawsuit over the Wagon Road Lands, now owned by Menasha

Deposit and Trust Company on March 8, 1884, to underwrite the bonds sold to support the company purchases in Coos Bay. The mortgage permitted that the majority of the bondholders could replace the Boston Safe Deposit and Trust Company as trustee, should they so choose, but they must provide that the minority holders remained secured. It is the writer's opinion that OSIC defaulted on their obligation to the Trust Company and William J. Rotch and Edward Mandell stepped forward with their majority of bond ownership and effectively took the Trust Company out of the picture. They then foreclosed on the mortgage in the Circuit Court of Oregon, Case No. 1344: "That there is due plaintiffs as Trustees under the mortgage indentures all real and personal property (of OSIC) in the complaint of $1,300,000 plus interest at 6%/annum. . . totaling $1,516,666.66."

The plaintiffs were awarded the decision to recover their costs through a sheriff's auction to be held on April 11, 1887, on the steps of the Coos County Courthouse in Coquille, Oregon.

On June 23, 1887, the OSIC's land—including the Wagon Road acreage—was transmitted by sheriff's decree and a master deed to William J. Rotch and William W. Crapo. The initial board meeting of the SOC on February 6, 1888, authorized Elijah Smith "to purchase all of the property held by Mssr. William W. Crapo and William J Rotch heretofore owned by the Oregon Southern Improvement Company as conveyed to them by a Master in Chancery under decree duly rendered by the Circuit Court of the United States for the District of Oregon."

The price of the purchase was $850,000 in cash and $650,000 in SOC stock. Through these transactions, the Southern Oregon Company acquired more than 100,000 acres of prime timberland in southern Oregon originally held by Coos Bay Wagon Road Company and Henry Luse.

When the Menasha Wooden Ware Company acquired all of the stock of the SOC between 1908 and 1911, these same lands became their temporary property. The Southern Oregon Company became a wholly-owned subsidiary of Menasha Wooden Ware Company when it was acquired. It was held as a wholly-owned subsidiary until the mid-1940s. The timberland ownership was "temporary" because the federal courts would take it away from Menasha in 1916.

But the Wagon Road lands would prove a mixed blessing. The SOC now owned property that had originally been part of the 1883 OSIC/Coos Bay Wagon Road Company deal, which took the once-public land away from its intended purpose—that was, to be sold to small land owners in claims not to exceed 160 acres—and put it into the hands of timbermen. To many, the purchase of the "Road" and its timberlands by the OSIC had been a clear violation of the original Wagon Road Act; now the SOC became a target of the resulting legal actions.

Thomas S. Minot brought the first lawsuit against the SOC on behalf of several disgruntled neighboring landowners along the Wagon Road itself. The complaint was filed in the US Circuit Court in Portland, Oregon. After a relatively short trial in 1902, the case was decided in favor of the SOC and supposedly cleared the way for a clean title. However, a group of unhappy landowners carried on the fight against the SOC, pooling their interest under what became known as the Nichols claim and threatening to take their case all the way to the United States Supreme Court. It was not until 1915 that the Supreme Court ruled, reversing the circuit court in a landmark decision that swept the Wagon Road lands into the infamous Oregon and California (O&C) Rail Road Grant. The court said in essence that the Wagon Road (and O&C) lands were improperly disposed of following the acquisition by the OSIC and held that the federal government needed to repossess the lands (called revestment) and compensate the current owner at $2.50 per acre, less the value the owner received during the interim from any land sales.

Between 1902 and 1915, the Wagon Road lands were in legal limbo both because of the landowners' legal actions and the interest of the legislature of the State of Oregon into the disposition of the lands.

In a 1907 letter to one of the Southern Oregon Company co-founders back east, Elijah Smith (then president of the SOC) told of his testimony before the Oregon State Senate committee in opposition to a bill introduced by Senator Malarkey to challenge the SOC's title to the Coos Bay Wagon Road lands. The bill would have forced the company to sell the lands to individuals for the price of $2.50 per acre in 160-acre parcels.

It is apparent from the testimony that as the SOC president went along, his temper rose. At one point he even accused the State of Oregon of robbing the present stockholders of the Southern Oregon Company of their lands. He recited a comment by Judge Bellinger (the sitting judge of the earlier United States Circuit Court decision) stating that "The court would have to forget all the law it ever knew or thought it knew before he would entertain their proposition [of denying title to the Wagon Road Company lands]." Smith further argued the absurdity of selling some of these lands in 160-acre parcels, as much of the land was inaccessible, burned over, and essentially worthless!

The Southern Oregon Company store in Empire, 1910. The store was destroyed by fire shortly after this photo was taken (Coos Art Museum, Victor C. West Collection)

He concluded, "I have endeavored to have the committee understand that the mischievous statements so widely circulated in the papers recently as to my owning the property personally, being a timber baron, monopolist, etc., was unwarranted and wholly untrue and for the purpose of creating a sentiment and building up a prejudice against [the Southern Oregon Company] to assist and facilitate their present attempt when the fact was that there were over 100 people interested in the company and property, whose rights, for reasons given, were vastly superior and not to be compared to the parties now buying and seeking to use the State of Oregon as a tool to accomplish what would be a practical robbery so far as our stockholders were concerned and of no benefit whatever to the state itself."

It is no wonder that Elijah Smith and the SOC were upset. From records of a 1902 timber cruise conducted on the company properties (100,678 acres, both the Luse and Coos Bay Wagon Road lands), the following is summarized:

SPECIES	VOLUME in board feet
Fir	1,690,178,000
Spruce	15,876,000
Hemlock	69,945,000
Cedar	71,897,000
Piling	90,380,000
Total	1,938,276,000

This Orchard Land and Timber Company operation at Divide, Oregon, in 1910 is an example of just how portable these railroad tie mills were. Remnants of these operations can be found today by the enormous piles of sawdust and rotting wooden slabs. (Douglas County Historical Museum)

At the going "stumpage" rate at the time of $1.50– $2.00 per thousand board feet, this volume would be worth close to $3,500,000—considerably more than what was paid and well worth fighting for!

E. B. Dean mill along the east side of Isthmus Slough (ca. 1900). This mill was purchased by Charles A. Smith in 1906 for the purpose of supplying lumber for the mill he built at Bunker Hill in 1908. It was rebuilt by C. A. Smith in 1910 after a fire destroyed it (Coos Art Museum, Victor C. West Collection)

Frederick Weyerhaeuser

William J. Conrad mill in Marshfield (Coos County Historical Museum)

The Oregon Export Lumber Company mill on Isthmus Slough (ca. 1919) originally the Millington sawmill, rebuilt by the Export Company (Courtesy of Gordon Ross)

Gifford Pinchot

James J. Hill

TIMBER BARONS

In researching the history of Menasha Wooden Ware Company in the Pacific Northwest, one is struck by how Menasha and C. R. Smith connected with many of the other "timber barons" that moved west about the same time. Names such as Henry Luse, Asa Meade Simpson (not to be confused with Sol Simpson of the Simpson Timber Company of Seattle, Washington), Louis Simpson, Elijah Smith, Prosper W. Smith, James Hill, William Crapo, Charles Axel Smith, Al Powers, and Harvey MacMillan can be found in association with deals made with Menasha and its land agents in Oregon, Idaho, and Washington. Others who came west looking for timber along with C. R. included such notables as: Andrew Pope, William Talbot, Cyrus Walker, Simon Benson, Thomas D. Stimson, and Frederick Weyerhaeuser. While the record is silent as to their direct involvement with Menasha, one can surmise that indeed they did "know" each other.

Many documents detail business transactions between Menasha and other timber companies in the region at the time Menasha settled in Coos Bay, including the Simpson Lumber Company, Weyerhaeuser Timber Company, C. A. Smith Lumber Company, Smith-Powers Logging Company, Coos Bay Lumber Company, Northern Pacific Railroad, North Bend Lumber Company, and the Southern Oregon Company. Each transaction would undoubtedly reflect the colorful stories of the era. However, such anecdotes are lost to the dust of time.

Many of these people and companies got their start in America's Midwest. As lumbermen liquidated the pine forests from Wisconsin to Minnesota, they began to turn their attention and financial resources to the extensive forests of the Pacific Northwest. By the 1870s, logging and lumbering in Coos County were showing signs of expanding rapidly. Towns like Empire City (now Empire), Yarrow (now North Bend), and Marshfield (now Coos Bay) grew rapidly around the large mills built by Asa Simpson, Henry Luse, and E. B. Dean.

Several other sawmill operators tried their hand at making lumber in competition with these big mills, but many of them failed to gain the economies of scale that these three competitors possessed. Small mills—like the one located on the flats at Millington currently occupied by the Knutson Company Log yard—had little chance to compete with Dean, Simpson, and Luse.

Other "gyppo" mills cropped up closer to the forest to provide a ready outlet for the small loggers and farmers who did not have the means of transporting their logs great distances. These mills were simply too small to create a workforce that might lead to a permanent community, although there are exceptions: the small town of Broadbent, Oregon, was created in this way.

The Cedar Products mill in Broadbent, Oregon, located more than thirty miles from Coos Bay (ca. 1921) (Coos County Historical Museum)

And that was just the value of the timber. The SOC also owned the Big Mill in Empire City and adjacent structures, including a large adjacent wharf supported by 4,000 white cedar pilings with about 700 feet of frontage on the main steamer channel, a two-story office building with a fireproof vault, and a company store 30 feet x 100 feet in dimension. The Big Mill was idled for some time around the turn of the century due to depressed lumber markets in California, but the company store continued to operate as a stand-alone profit center and reportedly had on hand some $14,000 of goods for sale. The initial deliveries to the Empire store were made by the company's ship *AL-KI* as company records indicate that the local merchants apparently "charged excessive profits for their material."

The company also owned a large, two-story hotel (the Arago) in Empire City, as well as a stable, cannery buildings, a warehouse, and three residences (the author believes that Norm Laird, a Menasha timber cruiser in the 1940s and 1950s, occupied one of these dwellings as late as the early 1970s. It was eventually burned down by the local fire department as a hazard. The Empire City store was destroyed by fire in early part of the 1890s). In addition to the large freighter *AL-KI*, the SOC also owned the 79-foot sternwheel steamer *Montesano*, purchased in 1888 and used primarily for transportation of passengers and freight up to Catching Slough.

Moving two veneer blocks in the early twentieth century (Pacific County Historical Society)

The Southern Oregon Company also owned considerable developable land in and around Empire City. In addition to the mill site of about fifteen acres, the company store located on lot 8 in the Empire City subdivision, and the Arago Hotel on lots adjoining the store, the company owned 128 "valuable lots" within the original townsite of Empire City as well as 980 residential lots in townsite additions.

And there was more. In connection with its logging business, the SOC owned a "well graded and substantially built" railroad at Sumner, Oregon. The timber in the vicinity had all been cut by 1902, but there was still more than six miles of 40-pound track in inventory along with two locomotives (ten and eight tons in size), thirty-two logging trucks, a pump car, two flatcars, and associated stock. From SOC correspondence, it appears that Louis Simpson had used the rail line to haul out some of his timber in the area and did not pay the freight charge.

The forests around Coos Bay also contained many Cascara trees, whose bark could be used to make a laxative. This "Chittum bark" as it was known locally, was another potential source of income. A 1902 report in the Menasha files refers to:

> *Chittum Bark: This bark, or drug known as. . . "Cascara sagrada" or Chittum Bark is peculiar to [these] localities. The company's land near Sumner and Fairview and East Fork produce an immense quantity of the article. Market varies. At one time the price was 20 cents per pound. Present quotation is 3 cents per pound, delivered to San Francisco. At one time the company carried over 150 tons of this bark in stock, which amount was obtained during one summer, chiefly from company lands.*

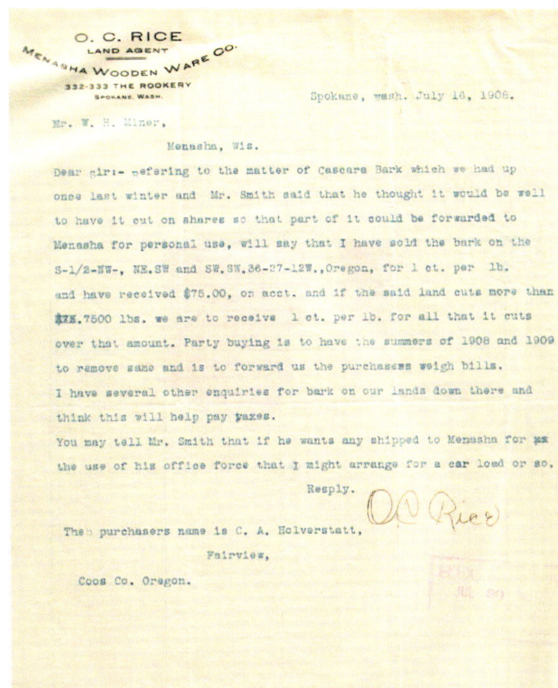

O. C. RICE
LAND AGENT
MENASHA WOODEN WARE CO.
332-333 THE ROOKERY
SPOKANE, WASH.

Spokane, wash. July 16, 1908.

Mr. W. H. Miner,

Menasha, Wis.

Dear Sir:- Refering to the matter of Cascara Bark which we had up once last winter and Mr. Smith said that he thought it would be well to have it cut on shares so that part of it could be forwarded to Menasha for personal use, will say that I have sold the bark on the S-1/2-NW-, NE.SW and SW.SW.36-27-12W.,Oregon, for 1 ct. per lb. and have received $75.00, on acct. and if the said land cuts more than 7500 lbs. we are to receive 1 ct. per lb. for all that it cuts over that amount. Party buying is to have the summers of 1908 and 1909 to remove same and is to forward us the purchasers weigh bills. I have several other enquiries for bark on our lands down there and think this will help pay taxes. You may tell Mr. Smith that if he wants any shipped to Menasha for the use of his office force that I might arrange for a car load or so.

Resply.
O C Rice

The purchasers name is C. A. Holverstatt,
Fairview,
Coos Co. Oregon.

O. C. Rice, land agent for Menasha, writes to Willis H. Miner about "Chittum bark" as apparently C. R. Smith had wanted some sent to him in Wisconsin.

Following the disappointments with their big mill in Empire City (which was closed in 1892) and the uncertainty surrounding the political battles over the Wagon Road lands, the SOC began trying to sell their timberland assets in the region. In 1905, the local newspaper reported that the SOC was negotiating to sell 100,678 acres to Frank B. White. No deal was struck, and in December 1906, the SOC's Elijah Smith was negotiating with C. A. Smith, the Swedish emigrant

Log bucker with "misery whip", a cross-cut log bucking saw (Coos County Historical Museum)

timberman introduced in Chapter 2, who was interested in moving west after building one of the largest sawmills in Minnesota. Apparently Elijah Smith gave C. A. Smith a sixty-day option to review the property for a fee of $16,000, which in the end was returned, as C. A.'s attorneys could not prove a clear title to the land.

The editor of the *Coos Bay Harbor* newspaper weighed in on the ongoing legal battles over the Wagon Road lands in February 1907, writing, "The *Harbor* makes the prediction that the Southern Oregon Company's title will hold and that nothing will result from the proposed litigation but to retard the development of Coos County. The business interests of the county should frown upon the jumping of the property. Had not this lawsuit threatened, the Southern Oregon property would have been bought by the C.A. Smith people who would have made extensive improvements of vast benefits to Coos County."

Meanwhile, the SOC continued its efforts to sell the Wagon Road lands. Since Menasha began purchasing timberland in the Coos Bay area in 1905, their local representatives were undoubtedly aware of the activities of the SOC. The record is lost as to whether C. R. Smith or other Menasha executives ever became acquainted with Elijah Smith or the other original subscribers of the SOC or its predecessor. While Elijah Smith appears to have been a very private man, he seemed to be held in high esteem by the local press as reported in numerous articles in the early 1900s. He died in Portland at age 72 on April 11, 1912, after a lengthy illness.

THE SOUTHERN OREGON COMPANY

AUTHORIZED CAPITAL STOCK $1,500,000. SHARES $100 EACH.

This Certifies that A.V. Wickman is the owner of One (1) Shares of One Hundred Dollars each of the full paid Capital Stock of THE SOUTHERN OREGON COMPANY, transferable only on the Books of the Company in person or by Attorney, on surrender of this Certificate. Witness the signatures of the President and Assistant Secretary of the Company, and its Corporate Seal affixed at Boston, Mass. this 20th day of July 1911

Here C. R. Smith signs as president of the Southern Oregon Company on July 20, 1911

Sometime around 1908, C. R. Smith decided to take on the risk of buying the Wagon Road lands. Between 1908 and 1911, the Menasha Wooden Ware Company acquired the stock of the SOC with many of the assets listed above still in place. While the records are scarce, the end-result was clear. In 1911 an announcement in the *Timberman* magazine stated:

> *The controlling interest of the Southern Oregon Co., owning about 100,000 acres in Coos, Curry and Douglas counties, Oregon, and a sawmill at Empire, Oregon, has been secured by the Menasha Woodenware Co., of Menasha, Wis. At the annual meeting recently held at Empire, the following officers were elected: President, C. R. Smith, of the Menasha Woodenware Co.; vice president, Herbert Armstrong, Western representative of the Menasha Woodenware Co.; secretary, Claude Hockett, who was with the Southern Oregon Company at Empire. Mr. Armstrong has assumed active control of the extensive holdings of the Menasha Woodenware Co. in Southwestern Oregon, with offices at North Bend. Mr. Hockett will continue in charge of the local affairs of the Southern Oregon Co. at Empire.*

THE SOUTHERN OREGON COMPANY

AUTHORIZED CAPITAL STOCK $1,500,000. SHARES $100 EACH.

This Certifies that William W. Crapo is the owner of Three hundred and sixteen Shares of One Hundred Dollars each of the full paid Capital Stock of THE SOUTHERN OREGON COMPANY, transferable only on the Books of the Company in person or by Attorney, on surrender of this Certificate. Witness the signatures of the President and Assistant Secretary of the Company, and its Corporate Seal affixed at Boston, Mass. this First day of July 1902.

P.W. Smith Elijah Smith

In July, 1902, the Southern Oregon Company officers were Elijah Smith, president and Prosper W. Smith, secretary/treasurer

A 1911 letter from William W. Crapo—a New Bedford, Massachusetts, attorney and a founder of the SOC—to C. R. Smith outlines an issue regarding back property taxes. Obviously, there was some connection between the parties, but to what extent is unknown. The SOC continued to operate as a wholly owned subsidiary of Menasha for another thirty-one years, dissolving into the parent in 1942.

Charles Robinson Smith, while on his way to the Pacific Coast in March 1916, suffered a cerebral accident on a train near Spokane, Washington. Mrs. Smith (Isabel Rogers) was summoned and brought her husband back to their New York residence. C. R. lingered in poor health and finally passed away on May 12,1916; he was sixty-one years old.

Menasha faced some grave prospects for the continuance of the family company after more than six decades of strong leadership under Elisha and C.R.

After C. R.'s death, three non-family presidents lead the company:

F. D. Lake 1916–1918

Thomas Kearney 1918–1921

Willis Miner 1921–1936

C. R. had three children, Mowry (Sr.), Carlton R., and Sylvia, but none were really ready to assume the leadership role of the company. It would not be until some twenty years later that a family member would resume the helm of Menasha when, in 1936, Carlton Smith was elected president of the Menasha Wooden Ware Company and his brother Mowry (Sr.) was elected president of the Menasha Wooden Ware Corporation.

Throughout the period between 1916 and 1936, the Menasha minute books contain many references to sales and potential sales of various properties. For instance, from a meeting of the Menasha board of directors, held at Menasha, Wisconsin, on September 26, 1916:

> *Moved by F. D. Lake the Southern Oregon Company negotiate sale through Herbert Armstrong to sell all the SOC land holdings in Coos County Oregon outside the Wagon Grant at a price of $1,000,000. Second by Mowry Smith Sr. . . . Motion carried.*

Another example is contained in a resolution of the Menasha board on November 14, 1917:

> *WHEREAS, Herbert Armstrong, president of the Southern Oregon Company has wired us that apparently he has an opportunity to sell [the former Big Mill in Empire City] and town site of Empire and the timber in Twp. 25, Rge. 13 W and Twp. 27 Rge. 14 W, without the land, belonging to the Southern Oregon Company and suggests making a price on the same of $850,000. . . . Motion carried*

Menasha must have also been extremely frustrated by the mess caused by the US Supreme Court decision regarding its Wagon Road lands. From the minutes of a meeting of the board of directors on January 8, 1918:

> *On motion of W. H. Miner, resolved that the action of the officers of the Menasha W. W. Co. in recommending to the officers of the Southern Oregon Company that if possible they settle the government suit on the [Coos Bay Wagon Road] Grant Lands at $2.50 per acre net to the Southern Oregon Company be ratified and approved. Motion seconded by F. D. Fish. . . . Motion carried.*

In fact the Supreme Court's decision to return the O&C and Wagon Road lands to the public had led to another problem: Who was going to buy the land from the companies that owned it? Eventually, the buyer would be the US government. In a special report to the shareholders on February 17, 1919, Menasha president Thomas Kearney wrote, "The Menasha Company is the

Thomas Kearney,
President of Menasha
1918–1921

Willis H. Miner,
President of Menasha
1921–1936

Mowry Smith Sr.

This stock certificate issued to Henry S. Smith, brother of C. R. Smith, on June 7, 1911, suggests that Elijah Smith was still president of the Southern Oregon Company

Log jam on the Coquille River (ca.1907). The logs were destined for the Prosper Mill near Randolph (Courtesy R. Gess Smith)

The Prosper Sawmill on the Coquille River (ca. 1905) (Coos Art Museum, Victor C. West Collection)

Log raft in Bay (Coos County Historical Museum)

BOOM TIMES

In the early days, water was the only real method of transporting logs the long distances from where they were cut to where they were milled. As timbermen worked the forests of New England and the upper Midwest, they generally used the area's numerous lakes and bays to store their logs in inventory until the mill was ready for them. In the West lakes were few and far between, and often not strategically located for mill log inventory storage. The early sawmills therefore tended to be built adjacent to tidewater, often miles from their logging camps. In these cases the loggers would utilize major rivers or secondary streams to "splash" the logs to open water.

Logs from these early harvests were "yarded" downhill and dumped into backwater holding ponds in the local rivers. The logs would then be transported down river during winter freshets that generated sufficient water flow to float the logs to the sawmills. This meant the log drives had to be coordinated with winter weather, often leaving the mills without adequate logs during the dry summer months. Soon after the turn of the century, all the timber close to Coos Bay mills had been harvested, winter freshets began to be used to drive logs longer distances, and the era of "river drives" started on the South Fork of the Coos River in 1906.

Regardless of the method of dragging the would-be lumber away from the stump, the logs would eventually end up in the water for storage and later delivery to the mill. Even today, logs are stored in the backwater sloughs of the Coos Bay estuary.

As the logs were dumped into the water, men with long "pike" poles would arrange them in a parallel position to facilitate towing. When Menasha had its plywood operation in North Bend during the 1950s and 1960s, the same methods were employed. This was dangerous business as the logs would move apart and the unwary boom man might slip and fall in the water, the logs closing in over his head. Drowning during the movement of logs in a raft was not uncommon.

principal stockholder of The Southern Oregon Company. . . . The carrying out of the plan of settlement necessitated the adoption by Congress of a Bill appropriating a sum of money to the Oregon Company. Considerable delay was experienced in the passage of the Bill and I have assisted Mr. W. H. Miner in securing a final vote in the House and the Senate on this bill. . . . When the money is paid, it will enable the Southern Oregon Company to pay some of its obligations to the Menasha Company."

The final outcome of this fifty-year argument over the method of disposing of the government grant lands in Oregon would bring them back into public ownership through a revestment of the O&C lands (2,891,000 acres) and the Coos Bay Wagon Road lands (about 93,000 acres) for a gross payment of $2.50 per acre. Today the Bureau of Land Management manages these lands for the people of the United States.

At the January 21, 1919, meeting of the directors of the Southern Oregon Company in North Bend, Oregon, a motion was made John G. Mullen and unanimously approved to adopt the following resolution:

> *Resolved that this company hereby consents to and accepts the provisions of the Act of Congress approved February 26, 1919 entitled: "An Act to accept from the Southern Oregon Company, a corporation organized under the laws of the State of Oregon, a reconveyance of the lands granted to the State of Oregon by an Act approved March 3, 1869, entitled "An act granting lands to the State of Oregon to aid in the construction of a military wagon road from the navigable waters of Coos Bay to Roseburg, in said State commonly known as the Coos Bay Wagon Road Grant to provide for the disposition of said lands and for other purposes; and that pursuant thereto the President and the Secretary of this Company be and they are authorized and empowered for and on behalf of the Company and in its name and as its act and deed to execute and deliver a deed reconveying to the United States of America all of this Company's right, title and interest in and to all the lands granted to the State of Oregon by the act of Congress approved March 3, 1869, entitled "an Act granting lands to the State of Oregon to aid in the construction of a military wagon road from the navigable waters of Coos Bay to Roseburg in said State", and which by authority of Congress and the legislature of the State of Oregon were included in Patents numbered 1, 2, 3 and 4 issued February 12, 1875, November 18, 1876, March 18, 1876 and February 17, 1877, respectively by the United States to the Coos Bay Wagon Road Company from which latter company this Company acquired title through means conveyances, except so much of said lands not exceeding 12,057.24 acres which have heretofore been sold to other parties by this Company and its predecessors in interest.*

It becomes painfully obvious from reading the record that the founders of both the OSIC and SOC never realized their dreams of creating an Empire City-based rail-coal-lumber-shipping empire along the shores of Coos Bay. They also probably lost millions of dollars on the venture.

A resolution was adopted by the shareholders of Southern Oregon Company (Menasha Wooden Ware Company owned 98.8 percent of the company) to dissolve the SOC on May 18, 1942.

The Brookings Timber and Logging Company portable sawmill used in Curry County for cutting railroad timbers and ties (Douglas County Historical Museum)

The Smith-Powers Logging Company steam shovel building a rail grade near Powers, Oregon (early 1900s) (Coos County Historical Museum)

"HELL FOR STOUT"

While Henry Luse and Asa Simpson had introduced steam power to their Coos Bay sawmills in the 1850s, it would be some forty years before steam was adapted to the needs of Oregon logging crews.

With the advent of steam "donkeys" (a term affectionately used to describe the steam powered mechanical cable and spool systems used to drag logs along "skid roads" to a central location) the logs could be dragged great distances along skid roads in the same manner used by the old bull teams. John Dolbeer is credited with converting steam power for use in the logging operations out west in 1881. These steam engines generated considerable power and the cotton ropes used in the early days would constantly break. With the invention of the "wire rope" in 1890, the steam donkey found a real niche in the logging business.

C. A. Smith and his partner Al Powers were among the first to bring steam-powered logging to the region. A documentary film made in 1929 noted that the Coos Bay Lumber Company (a successor to C. A. Smith's company) had fifty steam donkeys and a dozen locomotives working in Coos County. Since Menasha would not directly enter the logging business until 1954, the early years of logging on company lands around Coos Bay generally involved selling standing timber tracts to local mill operators. In turn, they used their steam powered logging equipment to harvest the Menasha timber.

Donkeys were not the only form of steam power. At the turn of the century railroads became more commonly used in logging the backcountry of Coos, Douglas, and Curry counties. However, they were expensive to build and limited by the terrain they could traverse. When the "road" had to cross deep ravines, elaborate bridges were constructed.

As steam power began working its way into the woods of southern Oregon, rail lines were built deeper into the forest. Where grades in the rugged hills were too steep for the steam engines to pull the empty cars, ingenious mechanisms were designed to "yard" the cars uphill. A reverse mechanism was employed to lower the fully loaded log car back down the grade to reconnect with the engine for the trip to the water's edge.

Without a doubt, the railroad bridges used for reaching deep into the forests of the Pacific Northwest were built "hell for stout" using both sturdy Douglas-fir timbers and ingenious engineering. But on occasion, the bridge collapsed under the heavy weight of the locomotives and log-filled rail cars.

Where hillsides were steep, steam shovels mounted on a rail car would carve the mountains ahead of the track layers.

Shortly after the Great Depression started in 1929 the nation's demand for railroad ties exploded as rail spurs and mainlines were built nationwide as well as in Coos County. The nation was on the move and the railroads provided the vehicle of choice for moving cargo from point to point. As a consequence, "tie mills" sprang up in the woods throughout Oregon and Washington. These portable mills were an efficient way of cutting railroad ties as the tolerance on the dimensions of the ties was inexact; cutting them did not require the investment needed for a stationary sawmill. Many of these Gyppo tie mills were no more than a rickety platform, a sliding carriage, and a stationary circular saw with power supplied by a rudimentary steam engine. Often loggers would purchase a tract of timber, move their "mill" to the logging site, and begin skidding logs directly to the mill. The site where these tie mills were located were later evidenced by the huge piles of slabs cut from the second growth small timber as the sawyer squared up the log.

When the Tacoma Eastern Railroad reached the town of Morton, Washington, the community quickly grew to become the railroad tie manufacturing capital of the United States. At any one time during the 1930s to 1940s, there might be a mile of ties stacked eight to ten feet high awaiting the next shipment on the Tacoma Eastern.

As bull teams gave way to steam donkeys and later diesel powered yarders and tractors, and the rail lines could only reach specific areas where sufficient volume of timber could efficiently be skidded to the rail head, ingenious (and sometimes humorous) log trucking systems were devised. The most common were "corduroy roads" where planks or small logs were laid perpendicular to the roadway to provide traction.

This photo of a steam donkey shows the main line (on the left) used for pulling the logs to the landing and the haul-back line (the smaller cable on the right) used for dragging the main line back out into the woods for the next turn of logs to be skidded (Coos County Historical Society)

The steam engine was placed on logs hewed out of straight, defect-free Douglas-fir trees (University of Oregon)

Scranton narrow-gauge rail line log dump (Courtesy of Warren Dick)

A railroad bridge being constructed with a steam-powered pile driver (Douglas County Historical Museum)

Building a railroad trestle 1895 (Oregon Historical Society 62778)

High trestle, Coast Range Lumber Company (Douglas County Historical Museum)

As was the case in Oregon, teams of oxen were used to skid the logs out of the woods in Washington (University of Oregon)

chapter four

Expansion

During the first fifty years of the twentieth century Menasha purchased extensive timberlands in Oregon, Idaho, and Washington. During this era, it amassed 156,689 acres of western timberland at a total cost of $2,826,000. According to Menasha records, these lands during the same time period produced total income of $6,300,000 from land sales and $8,700,000 from timber sales. Obviously, these investments had paid off handsomely, and continue to pay exceptional dividends from the vision of C. R. Smith and the executives that followed him.

The push west by Menasha Wooden Ware Company in the continental United States was part of an obvious if unstated threefold strategy to:

- Ensure adequate supplies of lumber for Menasha's wooden ware factories in Wisconsin,
- Capture a share of the growing wooden ware business in the West, and
- Participate in the great western timberland opportunities, especially in the Pacific Northwest.

As the spruce and cedar lumber for Menasha butter tubs and barrels became more and more difficult to obtain from the Midwest forests, the vast timber stands of the Pacific Northwest provided an obvious solution.

To reach his goals, C. R. Smith and the company executives who followed him after his death on May 12, 1916, engineered an aggressive program of Menasha growth, diversification, and timberland purchases in a number of western states.

Oregon

Menasha began acquiring timberland in Oregon in 1905. In order to shield the seller from the fact that a large Midwest timber company was "in play" for the purchase, transactions were commonly made to land agents of the company—primarily O.C. Rice and Ole C. Sether in Oregon—who were used as "front men" for the purchases, along with C. R. Smith and, occasionally, Willis Miner. Both Rice and Sether had a prior connection with Menasha in the Midwest; their names also surface as land agents for Menasha in purchasing property in Minnesota and Michigan.

In an internal communiqué, Carlton Smith, younger son of C. R. Smith, recalled a story that was told by Willis Miner, who often accompanied C. R. on his westward trips:

> *It seems that Mr. Smith [C. R.] was discussing a prospective timber purchase with the two parties in question. It must have been some discussion because it led to a fistfight, leading Mr. Miner to conclude that the timber purchase would never be accomplished. But Charles R. was somehow able to get the quarrel settled, and was also able to make a successful bid on behalf of the company.*

Menasha owned vast stands of Sitka spruce in the Pacific Northwest, using the wood to make butter tubs (University of Oregon)

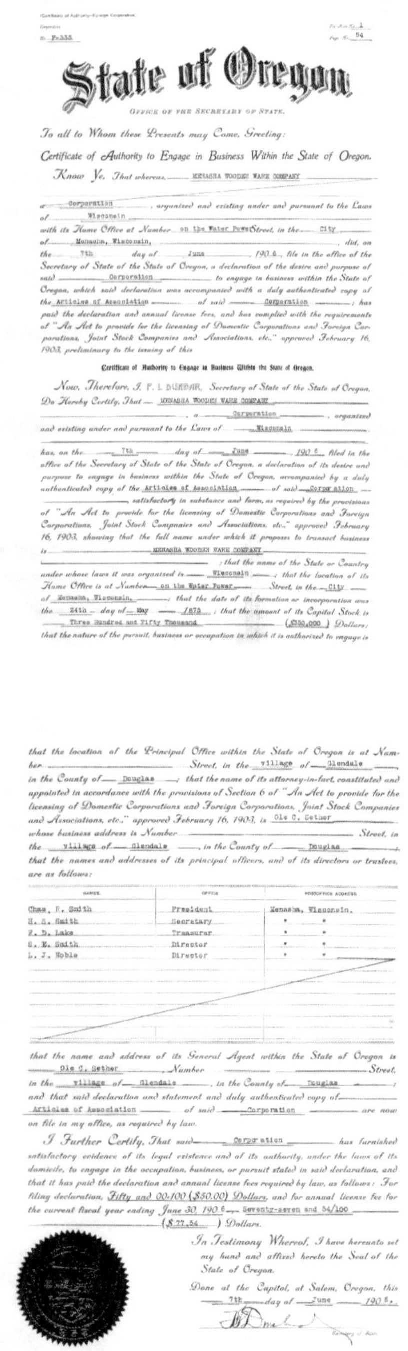

No. F-335

Fee Rec. No. 1
Page No. 54

State of Oregon

OFFICE OF THE SECRETARY OF STATE.

To all to Whom these Presents may Come, Greeting:

Certificate of Authority to Engage in Business Within the State of Oregon.

Know Ye, That whereas, MENASHA WOODEN WARE COMPANY a Corporation, *organized and existing under and pursuant to the Laws of* Wisconsin *with its Home Office at Number* on the Water Power *Street, in the* City *of* Menasha, Wisconsin, *did, on the* 7th *day of* June *190*6, *file in the office of the Secretary of State of the State of Oregon, a declaration of the desire and purpose of said* Corporation *to engage in business within the State of Oregon, which said declaration was accompanied with a duly authenticated copy of the* Articles of Association *of said* Corporation; *has paid the declaration and annual license fees, and has complied with the requirements of "An Act to provide for the licensing of Domestic Corporations and Foreign Corporations, Joint Stock Companies and Associations, etc.," approved February 16, 1903, preliminary to the issuing of this*

Certificate of Authority to Engage in Business Within the State of Oregon.

Now, Therefore, I, F. I. DUNBAR, *Secretary of State of the State of Oregon, Do Hereby Certify, That* MENASHA WOODEN WARE COMPANY, *a* Corporation, *organized and existing under and pursuant to the Laws of* Wisconsin *has, on the* 7th *day of* June *190*6 *filed in the office of the Secretary of State of the State of Oregon, a declaration of its desire and purpose to engage in business within the State of Oregon, accompanied by a duly authenticated copy of the* Articles of Association *of said* Corporation *satisfactory in substance and form, as required by the provisions of "An Act to provide for the licensing of Domestic Corporations and Foreign Corporations, Joint Stock Companies and Associations, etc.," approved February 16, 1903, showing that the full name under which it proposes to transact business is* MENASHA WOODEN WARE COMPANY; *that the name of the State or Country under whose laws it was organized is* Wisconsin; *that the location of its Home Office is at Number* on the Water Power *Street, in the* City *of* Menasha, Wisconsin; *that the date of its formation or incorporation was the* 24th *day of* May 1875; *that the amount of its Capital Stock is* Three Hundred and Fifty Thousand ($350,000) *Dollars; that the nature of the pursuit, business or occupation in which it is authorized to engage is*

that the location of the Principal Office within the State of Oregon is at Number ___ Street, in the village *of* Glendale, *in the County of* Douglas; *that the name of its attorney-in-fact, constituted and appointed in accordance with the provisions of Section 6 of "An Act to provide for the licensing of Domestic Corporations and Foreign Corporations, Joint Stock Companies and Associations, etc.," approved February 16, 1903, is* Ole C. Sether *whose business address is Number ___ Street, in the* village of Glendale, *in the County of* Douglas; *that the names and addresses of its principal officers, and of its directors or trustees, are as follows:*

NAMES	OFFICE	POSTOFFICE ADDRESS
Chas. R. Smith	President	Menasha, Wisconsin.
H. S. Smith	Secretary	" "
F. D. Lake	Treasurer	" "
S. M. Smith	Director	" "
L. J. Noble	Director	" "

that the name and address of its General Agent within the State of Oregon is Ole C. Sether, *Number ___ Street, in the* village *of* Glendale, *in the County of* Douglas; *and that said declaration and statement and duly authenticated copy of* Articles of Association *of said* Corporation *are now on file in my office, as required by law.*

I Further Certify, That said Corporation *has furnished satisfactory evidence of its legal existence and of its authority, under the laws of its domicile, to engage in the occupation, business, or pursuit stated in said declaration, and that it has paid the declaration and annual license fees required by law, as follows: For filing declaration, Fifty and 00/100 ($50.00) Dollars, and for annual license fee for the current fiscal year ending June 30, 190*6 Seventy-seven and 54/100 ($77.54) *Dollars.*

In Testimony Whereof, I have hereunto set my hand and affixed hereto the Seal of the State of Oregon.

Done at the Capitol, at Salem, Oregon, this 7th *day of* June *190*6.

In 1906, the Menasha Wooden Ware Company was officially recognized as an Oregon business

The first timberland purchased by Menasha in Oregon, as noted in Chapter One, was 318 acres bought for $2,200 on September 21, 1905. Three months later another 480 acres was acquired from the same seller for $5,760.

Rumors about timber locators, "entrymen," and land buyers—their pockets, reportedly, flush with greenbacks—filled the pages of Oregon newspapers at the time. The speculators and their friends in the public land offices played fast and loose with state and federal land laws to transfer thousands of square miles of timber to private ownership. According to a prominent lumber journal, many of the investors were "Great Lakes capitalists" who had exhausted the pine forests of that region. The Portland *Oregonian* estimated in 1902 that eastern capitalists had already invested $30 million in the Northwest. As the writers at the *Oregonian* put it, there "was a careful and complete exploitation of this country by men from Michigan and Wisconsin."

The art of successfully blocking up large acreages under a single ownership, according to an article in the *Oregonian*, involved "no one knowing what the buyers were up to." And in few places did the speculators work with greater skill than in southwestern Oregon.

Was the Menasha Wooden Ware Company's C. R. Smith one of those "eastern capitalists"? He certainly appeared to be by the size of his land purchases. Ole C. Sether, one of Menasha's land agents, purchased under his own name more than 10,000 acres in Oregon's Coos, Douglas, and Josephine counties for the company during the summer of 1906 alone.

As William G. Robbins of Oregon State University commented in a historical treatment of early timberland purchases in the region, "There is another and darker side of to those purchases [he names Menasha Wooden Ware Company, Charles Axel Smith, and Weyerhaeuser]: They assured that management and harvesting decisions would be made far from Coos Bay and that the 'economic colonialism' of the nineteenth century would continue."

Robbins' accusations about Weyerhaeuser and Menasha appear ill-founded. As time would tell, Menasha and Weyerhaeuser would remain an integral part of the economy of Coos Bay for the next 100 years.

On the other hand, C. A. Smith seemed to reach national prominence through using a land agent named Stephen Puter to accomplish some of these reportedly shady land deals in Oregon. Puter commemorated his escapades in his book *Looters of the Public Domain*, written in 1906 while he was spending seventeen months in Portland's Rocky Butte prison for what were apparently illegal land deals. However, regardless of the claims by Mr. Puter, C. A. Smith was also a real entrepreneur who built a huge sawmill in upper Coos Bay , helped push forward the deepening of the Coos Bay channel, and played a major role in developing the industry in southwestern Oregon.

There is no doubt that timbermen in the area used every method at their disposal to accumulate the real riches of the region. The vast forests were a real investment magnet. For instance in 1902 Frederick Weyerhaeuser and James J. Hill (*the* major player in the Northern Pacific Railroad) put their first "timberland deal" together when Weyerhaeuser's company purchased 900,000 acres of timberland in Washington from the Northern Pacific for $5,400,000.

The creation of the Weyerhaeuser Millicoma tree farm (which lies within the Millicoma River drainage east of Coos Bay) was connected to the creation of the Mount Rainier National Park, established by President Theodore Roosevelt in 1899. Much of what would become the Park's land was owned originally by the Northern Pacific Railroad and its owner, James J. Hill. To create the Park, the US Congress, under the provisions of the "Lieu Land Act," exchanged federal lands elsewhere for Northern Pacific holdings around the mountain. As one historian later put it, Congress traded "glacier lands owned by the Northern Pacific Railroad for prime Douglas-fir lands in Oregon and Washington."

Some of the land the Railroad obtained in the trade—31,000 acres in the Millicoma River Drainage—was purchased in July 1902 by Weyerhaeuser for $5 per acre. This was the first of many acquisitions by Weyerhaeuser in the area.

A story in the *Coos Bay Times* on February 26, 1908, headlined "Large Land Owners," printed "a list of the large land owners in Coos County, as compiled by the Coquille *Sentinel*. The list is a complete one and will prove interesting to our readers who desire to know the persons who own large tracts of land."

Carlton R. Smith inspecting an old-growth log on Menasha land in Coos County

OWNER	ACRES
C. A. Smith	38,589
C. A. Smith and others	10,202
Pillsbury Lumber Co.	22,387
Charles L. Trabert	13,209
Fred A. Kribs	1,909
Menasha Wooden Ware Co.	15,494
Weyerhaeuser Co.	27,075
Simpson Lumber Co.*	26,251
Southern Oregon Co.*	81,996
Oregon and Cal. Railroad Co.	102,949
Coos Bay Lumber and Coal Co.	8,040
Southport Land Co.	3,960
Glasgow Coal Co.	2,205
Jessie Holway*	2,243
Merchant Land Co.	2,638
George Mandigo*	840
23 other owners	56,714
Total	416,801

**Owners of timberland who would sell all or part of their holdings to Menasha.*

The C. A. Smith mill—later owned by the Coos Bay Lumber Company and eventually Georgia-Pacific Corporation—was reportedly one of the nation's largest. In 1987 Menasha would lease a portion of the mill site and build a red cedar sawmill named CB Cedar (Courtesy M. Reusser Collection)

Early in the twentieth century Asa Meade Simpson, in order to meet certain debt obligations, sold several parcels of land around Coos Bay to Menasha. Many of these same parcels are still owned by the company today.

By 1914 Menasha owned more than 32,000 acres of timberland in southern Oregon, purchased at an average price of $23.35 per acre. During the first fifty years of the twentieth century, Menasha assembled 51,122 acres of timberland in southwestern Oregon for a total cost of $1.3 million. During the same fifty year period, Menasha sold 10,800 acres in Oregon for $774,000. In addition to the land sales, timber was sold to local mills totaling some $5,000,000.

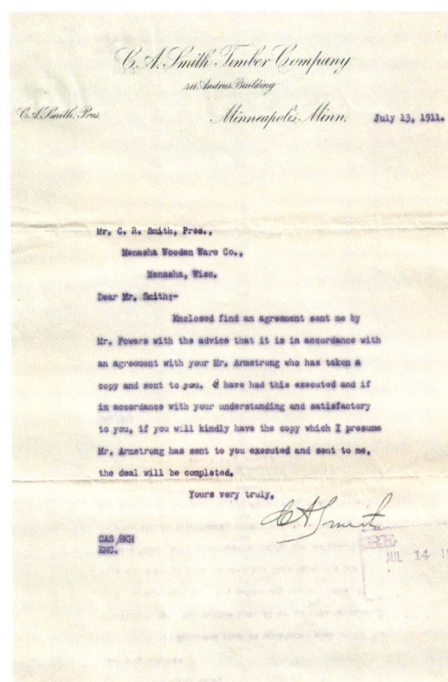

C. A. Smith Timber Company
Andrus Building
C. A. Smith, Pres.
Minneapolis, Minn. July 13, 1911.

Mr. C. R. Smith, Pres.,
Menasha Wooden Ware Co.,
Menasha, Wisc.

Dear Mr. Smith:-

Enclosed find an agreement sent me by Mr. Powers with the advice that it is in accordance with an agreement with your Mr. Armstrong who has taken a copy and sent to you. I have had this executed and if in accordance with your understanding and satisfactory to you, if you will kindly have the copy which I presume Mr. Armstrong has sent to you executed and sent to me, the deal will be completed.

Yours very truly,
C A Smith

CAS/RKH
ENC.

JUL 14 19

1911 letter from C. A. Smith to Menasha's C. R. Smith

Idaho

The initial purchases of Idaho timberland by Menasha began in 1900. As in Oregon, Menasha commonly used "land agents" to front many of their purchases in order to disguise the true acquirer of the land. Sellers would certainly have raised their prices had they known a timber company from the Midwest was behind the purchase. Charles McGuire was one such agent, and his name is found on several of the transition deeds, as were the names of Charles R. Smith, Willis H. Miner, and O. C. Rice. McGuire became the local representative of Menasha in Spokane, Washington.

The largest single timberland acquisition by the company in the United States during Menasha's 155-year history occurred on September 26, 1900, when the company purchased 34,000 acres in Idaho from the Northern Pacific Railroad (James J. Hill) for $54,488. Two years later the company purchased 7,700 acres from the State of Idaho for $85,561.

GMC log truck (Oregon Historical Society 464)

At a meeting of the Board of Directors of Menasha Wooden Ware Company on June 3, 1904, it was resolved to establish an office in the town of Coeur d'Alene, in Kootenai County, Idaho, with John P. Gray of Wallace, Idaho, designated as attorney in fact.

By 1912, Menasha had purchased more than 63,000 acres of land and timber rights in Idaho at a total cost of $800,000. Income from the Idaho purchases began to flow almost immediately as Menasha started "churning" the property through direct sales of land and timber. In 1903, Menasha sold 33,000 acres to multiple buyers for $180,000 after it had already sold off a timber cutting contract for $150,000.

Large red cedar trees were especially attractive to C. R. Smith; the wood was well-suited for making his company's wooden tubs (ca. 1900) (Lewis County Historical Museum)

It appears that Menasha moved through Idaho rather quickly on its way to Oregon. More than $700,000 was recouped from selling timber cutting rights on Menasha lands in Idaho between 1900 and 1940. It would be another fifty years before Menasha would venture back into Idaho for a timber investment.

Land sales accounted for another $3.1 million as Menasha exited Idaho. The largest single timberland sale in the company's history was made to the Blackwell Lumber Company of Coeur d'Alene in 1911. A total of 26,000 acres was sold to Blackwell for $2,259,236 or $86.62 per acre. The transaction was on contract; Menasha took a mortgage on the property and $1,000,000 worth of preferred stock in the Blackwell Lumber Company itself. The stock carried a coupon rate of 7 percent and was eventually redeemed.

In 1993, Menasha would take another "run" at the Idaho timber business by establishing an operation in Lewiston, Idaho. This effort focused on acquiring timber-harvesting rights from local landowners but not purchase of the land itself. Logs generated from the harvesting programs were marketed to local mills with the better-quality logs sorted and shipped to Japan.

The logs that were destined for the Japanese market were transported by barge from Lewiston down the Columbia River and off-loaded at Menasha log yard in St. Helens, Oregon, some 400 miles downstream.

From there, the logs were consolidated with other Menasha logs and loaded onboard ocean-going vessels for their 6,500 mile journey to Japan, where they would be sawn into lumber for the Japanese housing market.

While the initial business venture in Idaho by Menasha began in 1903 and lasted some forty years, the more recent venture lasted only seven years. The demise of the more recent Idaho project was primarily caused by the falloff in the overall Asian economy and hence a precipitous drop in demand for housing—and ultimately an unacceptable drop in the price of the Menasha logs. The operation was closed in 2000.

Washington

Menasha records indicate that the company's first timberland purchases in Washington were made around 1915, after acquisitions were curtailed in Oregon. This was about the time Menasha finally "threw in the towel" and out of frustration sold its Coos Bay Wagon Road lands back to the federal government. When it came to making more investments in Oregon the mood was somber and the Washington region seemed to offer new opportunities for the growing company. Between 1914 and 1930, 22,303 acres of timberland were purchased throughout the western part of Washington (known today as "west-side timber") for a cost of $673,000—at prices ranging from $11 to $60 per acre.

Since most of the "management" for Menasha was centered in Coos Bay, the majority of the income from the Washington holdings came via direct land sales. From 1912 to 1950 a total of 7,232 acres were sold in Washington for $2,457,000 or $340 per acre. The largest single sale occurred in 1928 to Weyerhaeuser.

Menasha also began selling its Washington timber to local loggers and small mill operators. The sizes of the sales were relatively small compared to the Oregon operation, and the timber was usually sold separately from the land. From 1917 to 1950 timber sale income totaled $3,015,000. Undoubtedly many of these sales were made to railroad-tie sawmill owners.

Since most of the activity for Menasha during this period occurred in the Coos Bay region, only limited historical information on the Washington holdings can be found in company records. However, a letter written by Menasha's Washington manager Charles McGuire on September 5, 1917, describes a forest fire on Menasha timberlands in Lewis County that fall: "The fire started from the West Fork Logging Company operations in section 29, Twp. 13 S, Rge. 5 E. and burned in a southeasterly direction through sections 28, 33, 34 and the west half of section 2 Twp. 12 S, Rge. 5 E." The letter sounds a bit frantic as McGuire recites the challenges of fighting a wildfire with limited equipment. He goes on to say, "In this district there have been at least four to five hundred million board feet of timber burned. The Milwaukee Timber Co. and

Ed Ayers, logger for Menasha on Lewis County, Washington, timberlands during the early 1950s, was noted for his ability to log trees 100 feet in length to be manufactured into telephone poles

A bargeload of Menasha logs going through the locks on the Columbia River on their way to Menasha log yard at St. Helens, Oregon

Tittle Bros. logging on the Maimi River, Tillamook, Oregon (Oregon Historical Museum)

Weyerhaeuser Timber Co. and the Northern Pacific Railroad are the heaviest losers. The whole country is very dry and the fires will not go out until we have rain."

As is the custom even today, the person or persons responsible for starting the fire is liable for damages to the neighbor. McGuire approached the president of the West Fork Logging Company, upon whose operation the fire had started, and told him he would either have to purchase the timber burned or pay for damages done to it. They settled on a figure of 25 million board feet damaged by the fire, resulting in a payment to Menasha of $50,000.

In his annual report to the stockholders of Menasha in 1919, Thomas Kearney, then-president of the company, speaking about the Washington operations, reported, "The lands, timber and mining rights owned by this company in Washington are of great value." Two years later Willis Miner and Mowry Smith Sr. reported to the shareholders, "We should continue from time to time to make investments in Washington lands as long as timber can be bought at anything like the present prices [of $22–$25 per acre]."

Income continued to flow from these Washington timberlands over the next fifty years with the major logging activity occurring in the mid-1950s and 1990s.

Steam donkeys were used in logging operations in Washington before they were used in Oregon (Pacific County Historical Society)

In the early 1950s several "gyppo loggers" were harvesting timber from the Menasha timberlands. The Pries Brothers Logging Company harvested timber for Menasha throughout its Washington ownership in the early 1950s and, coincidentally, returned to some of the same units to log again in the mid-1980s

New Mexico

Menasha records covering activities in this region are sketchy at best. It appears that Menasha acquired a one-half interest in 18,000 acres of timberland in McKinley and Valencia counties in New Mexico in May 1913 from John Garvin and William MacFarlane for $93,750.

On July 23, 1919, Willis Miner, Menasha vice president, wrote in a report to the stockholders, "This purchase has been disappointing. There were not enough operators in timber in New Mexico to create any market for it. We have, however, sold off quite a large portion of the lands and the remainder represents an investment of $12,376. Some of the remaining lands carry iron ore in paying quantities. . . . There will, however, be no profit from this venture."

Company interest in the property was sold in 1928 for $9,289. It is unknown if there was any intermediate timber income from these properties prior to the sale.

Inside the North Western Wooden Ware plant in Tacoma, Washington (late 1920s)

Utah

Again the records of Menasha timber ownership in this region are limited. It seems that two parcels were purchased in Sevier County: one from Charles Zigler deeded to C. R. Smith in 1913 totaling 160 acres; and one from Isabel Johnson deeded to Mowry Smith Sr. and Carlton Smith in 1918. No other information is known about these parcels.

British Columbia

It appears from Menasha records that two timber licenses totaling 680 acres on Dakota Creek, B.C., were purchased around 1914 and belonged to Mowry Smith Sr. and his brother Carlton Smith. The cost of the licenses was $6,000 with annual fees of $154, which were paid every year until the licenses were signed over to others in 1947.

Menasha's Washington land manager, Charles McGuire, estimated there was 11.4 million board feet of timber on the licenses; but that it would be uneconomical to log it to salt water on Howe Sound and tow it to Vancouver.

Mowry Smith Sr.

Donald C. "Tad" Shepard, husband of C. R. Smith's daughter Sylvia Smith

New Leaders, New Markets

By the time C. R. Smith died in 1916, Menasha had achieved its aim of amassing substantial timberland ownership in the West, especially in Idaho and Oregon. While C. R. would not see all the benefits of his vision, he is credited with the foresight and passion that created Menasha's growth in timber.

After he passed away, as noted in Chapter Three, Menasha leadership was cast somewhat adrift. None of C. R.'s children were ready to assume the reins of the business. F. D. Lake immediately became Menasha's president, followed by Thomas Kearney in 1918, and then Willis Miner, who served as Menasha's president from 1921 until 1936 when Mowry Smith Sr., C. R.'s oldest son, became president. Mowry Smith Sr. would serve as president for the next twenty-five years.

In addition to continuing C. R.'s program of timberland acquisition, the new leadership also built new relationships with suppliers and explored new markets.

Around 1918 the company acquired an ownership interest in the North Western Wooden Ware Company, a Tacoma, Washington, operation that would supply Sitka Spruce staves and bottoms and cedar covers for use in Menasha's butter-tub products made in Wisconsin.

The North Western Wooden Ware Company was formed on March 29, 1902, by George P. Wright (who sold his interest three months later), H. M. Sarvant, and C. M. Riddell, who for $25,000 purchased a building on the Tacoma tideflats, a woodworking plant, and machinery that had been owned and operated by two other area companies.

At a meeting of the trustees of North Western Wooden Ware Company in February 1922, Mowry Smith Sr. was elected a trustee and president with a salary of $2,500 per year. Donald G. Turner was elected vice president with a salary of $1,200 per year. A meeting of the directors of Menasha on April 18, 1922, ratified the action of Mowry Smith Sr. to purchase controlling interest in this company.

Later, North Western Wooden Ware contracted with Menasha for the use of a patented wire sealing device. Menasha charged them a royalty on its patent of three cents per dozen pails and four cents per dozen tubs, and entered into a selling contract for certain kinds of tubs and pails. The relationship grew under the aggressive sales efforts of Donald Shepard Sr., the son-in-law of Charles R. Smith. When Menasha made the last butter tubs in 1942, this business was sold.

Plants like the North Western Wooden Ware operation generated a tremendous amount of wood waste in the form of chips and sawdust. Getting rid of it cost money. In the old days, this waste was either burned or dumped into landfills.

Ray Gamble, an entrepreneur from Tacoma, Washington, came up with a solution to the wood waste problem. In the late 1920s he approached Mowry Smith Sr. with a proposal to create a new business out of the residue from the North Western Wooden Ware plant. A partnership was created in 1929 and a plant was constructed in Tacoma to turn the waste into wood flour. Wood flour is a very fine inert wood dust material that would ultimately be used as a filler in many different products.

It was not long before Menasha bought out Gamble and a new company was formed: Menasha Tacoma Corporation. Leif Jensen became the local manager and Joe Diven became president.

As Mowry Smith Jr. recalls, "In the early 1950s, Jensen was over 75 years of age and the company needed younger management." Richard (Dick) Hansen was brought in to replace

The North Western Wooden Ware Company in Tacoma

Jensen as general manager in 1953 and expanded the business into six wood fiber plants scattered throughout Oregon and Washington: Eugene (1954); Albany (1963); Grants Pass (1967); North Bend (1955); Centralia (1974); and Marysville (1987). Shortly thereafter, the company name was changed to Menasha Wood Flour of Oregon, and finally on January 1, 1967, to the Wood Fibre Division of Menasha Corporation.

With the completion of the Grants Pass plant, Menasha became the largest single producer of wood flour in the United States. These plants produced Douglas-fir flour exclusively, using a grinding and mechanical gravity-sifting process to produce exact sizing specifications. A number of industries were served nationwide, including the phenolic molding material manufacturers, foundries, and also the makers of such things as explosives, compression molding, exterior finishes, plastic wood, pelting, and roofing materials.

Not only did Hansen expand the geographic distribution of the wood flour plants throughout the Northwest, he also expanded into a new, but related wood/bark-dust product: glue extender, used in the assembly of plywood. The two extender plants in Washington used Red Alder bark as their raw material and ground it into fine dust for mixing with the urea formaldehyde glues. A third extender plant in Albany, Oregon, used the sander-dust from the adjacent Willamette Industries particleboard plant.

In 1997, Menasha's Wood Fibre Division was sold to Ace International of Canada.

Richard (Dick) Hansen, General Manager, Menasha Wood Fibre Division

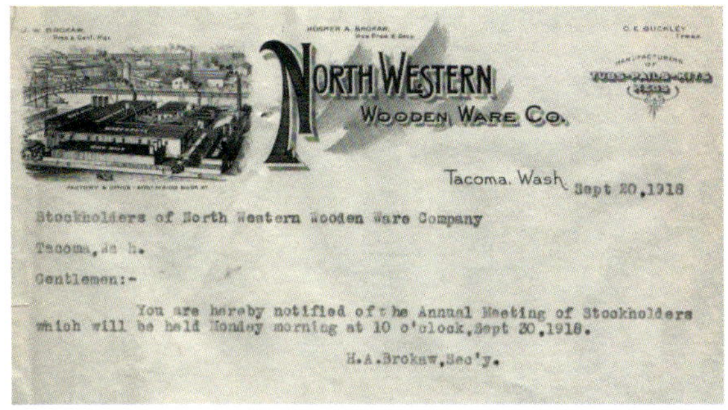

North Western Wooden Ware Co.

Tacoma. Wash. Sept 20,1918

Stockholders of North Western Wooden Ware Company

Tacoma, Wash.

Gentlemen:-

You are hereby notified of the Annual Meeting of Stockholders which will be held Monday morning at 10 o'clock, Sept 30,1918.

H.A. Brokaw, Sec'y.

Company records indicate that Menasha acquired an ownership interest in North Western Wooden Ware around 1918; Charles McGuire (Menasha's manager in Washington) was elected a trustee at that time

Menasha logs in a "flat raft" (single logs vs. bundles) just outside the in-feed conveyor to the plywood mill.

chapter
five

Plywood and Paper

The period 1950-1970 for Menasha was characterized by extensive expansion in both its timberland holdings in Oregon and the construction of major wood processing operations in the Coos Bay area. Aside from the Northwest Wooden Ware Company mill in Tacoma, Washington, Menasha had no wood manufacturing operation out West before 1950.

By 1950, Menasha owned in excess of 40,000 acres of timberland in southwestern Oregon that contained massive amounts of old-growth conifer timber. Some of this timber was of very high quality and some was quite defective. Many of the logs, however, would make excellent veneer for the rapidly developing plywood industry. The US housing market was booming, thanks to the return of World War II veterans. Plywood quickly replaced lumber as the base siding and roofing material for wood-framed houses.

In 1948 the Menasha Wooden Ware Corporation and Coos Head Timber Company entered into a fifty/fifty partnership called the Menasha-Coos Head Plywood Corporation. Their purpose was to construct a fully integrated new plant built on the Coos Bay waterfront in the town of North Bend that would produce Douglas-fir plywood for US markets from the logs harvested from Menasha timberlands in the region. Mowry Smith Jr. recalls that Willis Smith of Coos Head and Earl Weed, who managed the local timber assets of Menasha, were the impetus for this venture.

On July 1, 1948, the front page of the *Coos Bay Harbor* newspaper carried the following headline:

Coos Bay Harbor

VOL. XLII, NO. 27 NORTH BEND, OREGON THURSDAY, JULY 1, 1948 10c PER COPY

Big Plywood Operation Coming Here

NORTH BEND IS LOCATION FOR $2,000,000 FACTORY EMPLOYING 200 WORKERS

Construction will be started immediately on a $2,000,000 plywood plant to be located on the North Bend waterfront, it was announced today by the owner-operators, the Menasha-Coos Head Plywood Corporation. The plant will be located on the site occupied by the former Commando Packing Company building and property covering a distance of about a quarter of a mile south.

Construction of the Menasha-Coos Head Plywood plant began on a site previously occupied by the Commando Fish Packing operation. The Commando plant is the white building in the upper left corner of the photo

The Menasha plywood mill completed and ready for operation (ca. 1950)

This photo from the 1950s shows (l. to r.) Howard ("Lefty") Page, Willis Smith, and Chuck Duecy of the Menasha-Coos Head Plywood Corporation

The Menasha water tower at the north end of the plywood mill was used to supply fresh water for the steam plant and for fire safety. It stood like a sentry for twenty years until 1972 when it was torn down as a safety hazard. Here it displays a little holiday cheer

The Menasha plywood mill (ca. 1950)

The Menasha plywood mill (ca. 1960)

Al Stinchfield, manager of the Menasha plywood plant

Willis Smith and Charles M. Duecy had organized the Coos Head Timber Company on May 15, 1946. At the time of the merger, Coos Head operated a sawmill at Empire City, a pulp mill in Empire City, the Coos Head Builders Supply Company in North Bend, and was associated with the Elliott Bay Mill Company in operating the Myrtle Point Veneer Company at Norway, Oregon.

The making of plywood was not a new endeavor in 1949. In fact, the French had developed the general principle of plywood in the late 1700s, as did the Russians. The first US patent involving what would become "plywood" was issued December 26, 1865, to John K. Mayo of New York City. The patent states, "The invention consists in cementing or otherwise fastening together a number of these scales or sheets (veneers) with the grain of the successive pieces, or some of them running crosswise or diversely from that of the others."

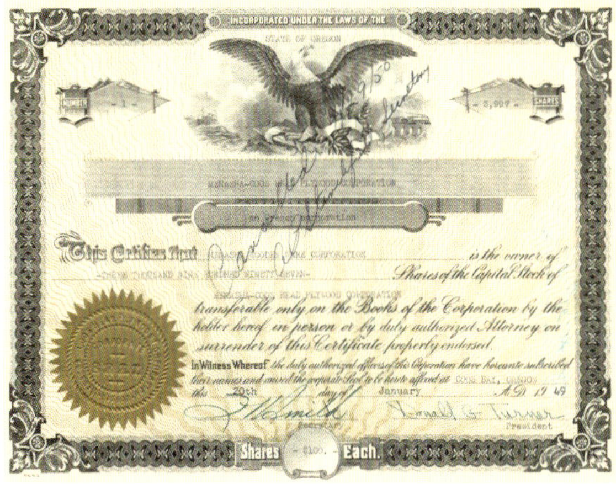

INCORPORATED UNDER THE LAWS OF THE

STATE OF OREGON

This Certifies that ... is the owner of ... Shares of the Capital Stock of ... transferable only on the Books of the Corporation by the holder hereof in person or by duly authorized Attorney on surrender of this Certificate properly endorsed.

In Witness Whereof the duly authorized officers of this Corporation have hereunto subscribed their names and caused the corporate seal to be hereto affixed at ... this 20th day of January A.D. 19 49

President

Shares ... Each.

The Menasha-Coos Head Plywood Corporation was chartered on June 22, 1948, with authorized capital of $800,000, which was all paid in, and constituted 8,000 shares at $100 par value. The capital stock of the new partnership was evenly divided between Coos Head Timber and Menasha

Robert Cour, in a treatise on the history of plywood, writes:

> *Sometime in March [1905] a panel crew was organized consisting of six men for the purpose of making plywood using an old St. Joe lathe and some excellent peeler logs. There was no press, no prepared glue, no clipper and no sanding machine. Their goal was to make plywood for the 1905 Lewis and Clark Exposition Fair in Portland. The men mixed animal glue and kept the very odoriferous material warm and pliable over a coal fire. The veneer was hung out on a covered loading dock and left to dry for two weeks. . . . A wooden press was improvised; the pressure was applied by putting ordinary house jacks between a backing timber and another timber fixed to the press. The work was slow with many failures, but they met their goal of having panels to show visitors to the fair.*

The agreement between Menasha and Coos Head Timber was simple: Menasha would supply the majority of logs for the new mill from its timber holdings in the area and Willis Smith, Chuck Duecy, and Lefty Page would run the plant. Coos Head would supply additional logs from their timberlands and logging operations. The combined workforce of the mill and the woods operations of both companies would exceed 700 men, according to Duecy.

The officers of the new Menasha-Coos Head Plywood Corporation were:

- Mowry Smith Sr. (Menasha), vice president
- Donald G. Turner (Menasha), president
- Earl R. Weed (Menasha), vice president
- F. Willis Smith (Coos Head), secretary
- Charles Duecy (Coos Head), manager

Coos Bay Pulp Corp. (University of Oregon)

Al Stinchfield, a CPA from the Moss-Adams accounting firm, was brought in as office manager in 1949 with the responsibility of developing new accounting practices for the operation.

On September 15, 1948, a preliminary construction budget was prepared for a total estimate of $1,559,740.10. It would employ a start-up crew of eighty people, growing to full strength employing 150 men with Richard D. Barrows as the general manager of the plant in 1949.

Carleton L. Smith, son of Mowry Smith Sr.

Mowry Smith Jr., son of Mowry Smith Sr.

At the time of the construction of the plywood mill, there were already forty-eight mills in the Pacific Northwest producing 1.8 billion square feet of Douglas-fir plywood. The Menasha-Coos Head operation would be entering an entrenched market and carving out a market share would prove a greater challenge than expected.

As with all the other mills on the Bay, the logs for the Menasha-Coos Head plant were hauled out of the water by means of a conveyor to the cut-off saw, where the logs were cut into 8-foot blocks, thence along a series of steam-operated rollers to await their "charging" into the lathe where the big logs were peeled to the required veneer thickness and grade. Steam furnished the power in almost all cases, because of its flexibility.

On March 29, 1950, Menasha agreed to buy out the 50 percent interest held by Coos Head Timber Company for $500,000 and formed the Menasha Plywood Corporation, a wholly-owned subsidiary of the Menasha Wooden Ware Corporation After the buyout, Chuck Duecy stayed on with Menasha to help with the plywood sales effort out of Chicago and Carleton L. Smith, youngest son of Mowry Smith Sr., took an active role in the mill. In 1950, Menasha's first full year in operation as a stand-alone mill, it shipped 34 million square feet of plywood and had a payroll of $890,000. Within five years production had risen to 60 million square feet of plywood with a payroll of $1,530,000.

Tragically, Carleton L. Smith died in 1952, as summarized in the Appleton *Post Crescent* obituary on April 4:

> *Carleton L. Smith, 31, sales manager of the Menasha Plywood Corporation, a subsidiary of the Menasha Wooden Ware Corporation, died last night at Coos Bay, Oregon, after an illness of ten days. . . . Carleton had worked for the Menasha Wooden Ware Corporation for about six years and was sent to Oregon two years ago. Carleton was the son of Mowry Smith Sr., president of the Menasha Wooden Ware Corporation.*

The Menasha management team at North Bend in 1965 (left to right) Boyd Swenson (chief forester); Bud Stebbins (controller); Homer White (general manager of Menasha Plywood); Dan (last name unknown); Harry May (purchasing agent); Al Stinchfield (manager of timberlands); Bill Haresnape (personnel manager); and Ralph Horn (timber manager)

Carleton L. Smith's brother Mowry Smith Jr. took over some of his duties, heading the sales division for the plywood plant. By 1955 the local management team included Allen P. Stinchfield, treasurer and plant manager, and Wilbur Craig, sales manager. Donald G. Turner remained as president of the company and Chuck Duecy handled the sales activity in Chicago.

Concurrent with the construction of the plywood mill in North Bend, the Weyerhaeuser Company built a massive mill complex (an old-growth sawmill and plywood mill) just up the Bay from the Menasha facility.

It is interesting to compare the scale of the Weyerhaeuser and Menasha operations between 1950 and 1970. Weyerhaeuser harvested some 10 billion board feet of timber from their ownership in the Coos River drainages during that period compared to 1.1 billion board feet harvested by Menasha.

In the mid-1950s, Menasha began supplementing the log supply for their plywood mill with timber purchased at auction from the Bureau of Land Management. For example, as the Coos Bay newspaper reported in 1951, "Menasha Plywood Corporation was the big buyer at the oral standing timber auction of the Bureau of Land Management Tuesday with a bid of $128,997 for 5 million board feet of good quality Douglas-fir (enough wood to build some 350 single family homes) timber in Coos County—outbidding the Irwin and Lyons Lumber Company." As fate would have it, some of these timber sales purchased by Menasha came from lands originally owned by the Southern Oregon Company that were revested back to the Federal government in 1916 by Menasha.

Bill Haresnape, the personnel manager for both the Menasha Plywood mill and later the Menasha paperboard mill in North Bend recalls how he went to work for Menasha in 1950:

> *I had come to North Bend from Nebraska to visit my brother who was stationed at the Navy base in Charleston [at the mouth of Coos Bay]. I did not have a job, but my wife and I rented a duplex on Broadway. The owner of the duplex asked if I had a job, and I told him no and did not plan on staying in North Bend very long. He was working at the new Menasha plywood plant and since I had never really seen this new product back home I said I would like a tour. I went to the mill, and before I knew it, my neighbor had introduced me to the personnel manager, Don Spencer, who had me fill out an application "just in case" I changed my mind and wanted to stay. Two days later, I was working as the day shift utility man. Eventually I became assistant personnel manager under Mowry Smith Jr.; and after Mowry returned to the Menasha headquarters in Wisconsin I became the personnel manager. I stayed with the mill until it closed in 1967 and became the personnel manager of the company's North Bend paper mill until I retired in 1997.*

In 1953, the Menasha Plywood Corporation began producing plastic-coated plywood. At that time, the product was primarily being used on concrete forms as the plastic would protect the wood and the sheet could be used over again. The sheet of plastic was laid on top of a sheet of plywood and inserted into the press for nineteen minutes. The record is silent as to whether this product was a success or not.

Menasha continued to invest in its North Bend plywood plant. On May 19, 1955, a *Coos Bay Times* front-page article read:

> MENASHA PLANS NEW MILL
> 14 MORE TO BE EMPLOYED
>
> Construction of a new modern two-by-four stud mill was announced by A. P. Stinchfield, general manager of Menasha Plywood Corporation. The mill will be located just south of the company's plywood plant. The stud mill is an important step in the maximum utilization of the company's timber resources and it is also tangible evidence of our confidence in the continued growth and prosperity of the Bay area. Start up is scheduled for August 15. The Menasha Sales Corporation, sales agent for the Menasha Plywood Corporation, will handle the lumber sales work.

In 1965, Menasha announced two expansions for the plywood operation: a 3,000 square-foot office building north of the water tower (and the current corporate office the Menasha Forest Products Corporation) and a renovation of the office space inside the plant to make way for a new sander operation.

By the mid-1960s the plywood market was in a slump. In January 1967 Menasha hired Dr. Barney Dowdle, an economics professor at the University of Washington, to evaluate the future prospects of their plywood business and its impact on Menasha timber prices. Dowdle's conclusions were:

1. Risk will increase for the plywood producers due to over-capacity;
2. However, plywood prices should increase from their current low point; and
3. Long-run timber prices will not be upward.

In hindsight, Professor Dowdle could not have been more wrong.

Within ten years plywood began to be replaced by oriented strand board and timber prices increased at compounded rates for the next thirty years. As Mowry Smith Jr. would often say later in his career, "Beware of experts."

After nineteen years of operation, Mowry Smith Jr. announced that Menasha's plywood and lumber operations would close permanently on September 1, 1967. Approximately 285 employees were affected by the closure, some being transferred to the pulp and paper mill across the Bay.

During its heyday in the 1950s and 1960s the Port of Coos Bay was known as the "world's largest forest products exporting port." There were:

- 8 large sawmills
- 2 large plywood plants
- 1 veneer and battery separator plant
- 1 pulp mill
- 1 shingle mill
- 1 small mill cutting alder
- Several small lumber and tie mills located in the woods
- 1 hardboard plant
- 1 wood flour plant utilizing wood waste

The Menasha paperboard mill can be seen at the left of this photo. A rail line adjoins the mill on the east side of the plant and a highway causeway can be seen running across the mudflats and connecting to Highway 101 just north of the highway bridge crossing Haynes Slough

Pulp and Paper

Plywood would not be the only new product to help Menasha profit from its western investments. In 1960, after several months of study, Mowry Smith Sr. announced the planned construction of a 100-ton per day capacity pulp and paper mill. "If plans go as expected," Smith wrote his employees, "the mill will be in operation in early 1961 with approximately 100 new employees added to the company payroll in the area."

The site selected for the mill was on Jordon Cove on Coos Bay. At the time of this announcement, the only access to the mill site was by water, rail, or a crude sand road along the west side of North Slough. It was soon realized that a causeway would have to be built to provide access to the site. Ernie Manders was chosen to head up the construction project for Menasha.

Several reasons were evident in the selection of this site for a mill (although an alternative site that was seriously considered was the Hinman ranch just east of Reedsport where the Roosevelt Elk Public Viewing Area is now located):

- The price was right.
- The wood supply was more than adequate and the chips would come from the red alder tree that essentially had no value at the time.
- There was already a large Menasha presence in North Bend with the existence of their plywood/stud mill, the wood flour plant, and the logging and booming operations purchased from Irwin and Lyons.
- An adequate water source could be developed from an aquifer under the extensive sand dunes adjacent to the proposed mill site. In 1961, the utility company estimated there was 152 billion gallons of fresh water contained in the 13,000-acre sand dune area, yielding 65 million gallons of water per day.

Coos Bay Pulp Corp. (University of Oregon)

Ernie Manders, construction manager of the North Bend paper mill, 1961

Several engineering challenges awaited Menasha:

- How much damage would be done to the equipment by the ever-blowing abrasive sand and salt air?
- Would the minerals in the water being pumped from the sand dune aquifer damage the quality of the paper?
- How would an access causeway be built across the tidal flats of the Bay from Highway 101 to the mill site?

In 1961, the leadership of Menasha was in transition. Mowry Smith Sr. was continuing as chairman of the board and Richard L. Johnson was elected president. Johnson was the one responsible for justifying the building the mill. As one insider remembers, "Dick could not make the economics work our for building the pulp and paper mill at North Bend, so he simply lowered the delivered wood cost estimate to the mill and the numbers miraculously became positive. The Board of Directors approved the investment ($5,000,000), the mill was built and it grew to be one of the best medium paper mills in the country. And it made tons of money."

Very little timber and few wood chips from Menasha timberlands were used by the mill. The mill was originally built to consume red alder chips, but Menasha was harvesting very little alder in its logging operations. Later the paper mill converted to using tan oak chips as its prime

A water well in a sand dune, just north of Menasha's paperboard mill, can be seen in the lower portion of the photograph. Pacific Power and Light developed the concept of extracting water out of the dunes in the late 1950s

Menasha's paperboard mill was in full operation by 1961. Note the planted Norway beach grass in the foreground. It was introduced into the region as a sand dune stabilizer for a town north of Coos Bay in the 1950s. The fast-growing grass, so successful at controlling the blowing sand, is now a problem to keep in check

The Menasha Board of Directors, 1964. Seated from left: Donald C. Shepard Jr., Richard L. Johnson, Tamblin Smith, Donald G. Turner, and Donald C. Shepard. Standing: Mowry Smith Jr., Roman Suess, and Oliver Smith

Richard L. Johnson, President of Menasha Corporation from 1961 to 1981

The North Bend paperboard mill building is shown under construction in 1960. Note the ubiquitous sand!

furnish (fiber) practically none of which was growing on Menasha timberlands. Still later the mill was converted to using an ever-increasing amount of recycled corrugated boxes as the main source of raw material until it became 100 percent converted to "old corrugated containers" (OCC) and the pulp mill side of the plant was dismantled.

The North Bend paper mill, along with the property known as the Henderson Ranch (originally purchased by Henry Heaton Luse in the 1850s, acquired by the Oregon Southern Improvement Company in 1883, then owned by the Southern Oregon Company, which eventually became a subsidiary of Menasha), a packaging plant in Anaheim, California, and a waste-paper recycling center in Portland, Oregon, were all sold to Weyerhaeuser in 1981. The transaction involved a complex exchange of stock from each company. The paper mill continued to run successfully for another twenty-three years until it was permanently closed by Weyerhaeuser in 2003.

The "wet end" of Menasha's 162-inch trim Black-Clawson paper machine. The Menasha paper mill was built to produce 125 tons of paper per day. By the time it was sold in 1981, improvements in engineering efficiency had raised output to 625 tons per day

A century and a half ago when Henry Luse purchased the Henderson Ranch property on the north side of Coos Bay, he probably did not realize its potential to become one of the best marine-related industrial parcels on the West Coast. The author spent several years positioning the abandoned ranch for some unforeseen industrial project. After several false starts, time would eventually prove the vision of OSIC to be correct: the lower Bay would indeed become a major industrial empire.

In 1962, the Menasha Wooden Ware Corporation changed its name to Menasha Corporation, as "wooden ware" no longer described the company's more wide-ranging operations.

With the closing of the Menasha "plywood/stud mill" in 1967, the closing of its logging operations in 1969, and the sale of the paper mill in 1981, all that was left of Menasha in North Bend in 1981 was its timberland holdings and one wood flour plant—a far cry from the major dominance it held in the 1950s and 1960s.

Still, without a doubt, the twenty years from 1950 to 1970 can be described as the period of greatest growth in forest products Menasha had known since the buildup of the woodenware business by Elisha Smith in Wisconsin in the late 1800s.

Towing logs on the lower portion of the Millicoma/Coos River (ca. 1960) (Courtesy of Warren Dick)

chapter six

A River of Trees

During the mid-1950s Menasha continued to invest heavily in its timberland ownership in Coos County. Two major acquisitions were made: the Irwin and Lyons Company purchase—which included 21,667 acres in the South Fork Coos River drainage, a logging camp at Tioga Creek, an extensive inventory of heavy-duty logging equipment, and the Coos River Boom Company—for $3,468,348; and the Al Peirce Company purchase of 11,948 acres west of Isthmus Slough and south of Coos Bay for $2 million.

The Irwin and Lyons acquisition was the largest single Oregon purchase made by Menasha up to that time (it would be surpassed only in 2003 with the purchase of 28,000 acres). It was precipitated by a tragic hunting accident that killed James Lyons, one of the owners of the company. The timberland portion of the acquisition, located in the upper reaches of the South Coos River drainage, as well as the timber itself was an ideal fit for the increasing demand for logs by the Menasha Plywood mill.

Mowry Smith Jr. remembered this about the purchase: "John Hawkins, the general manager of the Irwin and Lyons company had offered this timberland to Weyerhaeuser Timber Company for $3 million. They turned the deal down, and Mr. Hawkins came to the Menasha office in North Bend to present the same deal to Mowry Smith Sr. [Menasha foresters had been evaluating the property for some time and had a pretty good idea as to its potential]. My father asked that Mr. Hawkins step into the hall while they discussed the opportunity. A half-hour passed and Hawkins was called back into the room whereupon my father said, 'We'll take it!'"

According to Menasha records, the following allocation of the purchase price was made in 1954:

Land 21,667 acres	$65,000
Timber 183 MMBF	$1,743,348
Coos River Road	$1,000,000
Logging Equipment	$435,000
Coos River Boom Company	$125,000

Not only did Menasha acquire a large amount of old-growth timber and the Coos River Boom Company, they also bought the logging operation and logging camp of Irwin and Lyons. The logging camp was at the mouth of Tioga Creek, twenty-five miles above tidewater on the Coos River.

Bill Haresnape, personnel manager for the Menasha plywood mill, was also responsible for labor relations with Menasha logging division. Bill recalls that after the Irwin and Lyons purchase, there were two separate logging crews working the woods for Menasha: one that lived at the Tioga camp during the week with Les Fruin as foreman; another that traveled from town every

John Hawkins (left) was general manager of the Irwin and Lyons Company. Jim Lyons (right). They are both standing in front of the Mill B office

Howard Irwin (ca. 1953) (Courtesy of Jim Lyons)

The Menasha logging camp at Tioga Creek (ca. 1960) (Courtesy of Warren Dick)

The Menasha logging camp at Tioga Creek had a two-story bunkhouse that could accommodate forty men (Courtesy of Warren Dick)

The Menasha logging camp bunkhouse fire in 1962 (Courtesy of Warren Dick)

The most important part of any logging camp is the kitchen. Here, the Menasha camp cooks prepared the meals for the forty-plus men who lived at the camp during the week (Courtesy of Warren Dick)

day to work in the Mink Creek drainage with Ralph Mathews as the "side rod." The men from the town crew would be picked up early in the morning by a company "crummy" (a small bus built to withstand the rough logging roads and caulk boots of the loggers) for the two-hour drive each way to the work site. They often would not return home until well after dark.

As one story goes, Haresnape had a standing agreement with the local police departments regarding the payment of fines for the Monday morning release of Menasha loggers who had come to town on Friday night with a week's pay in their pockets and ready for fun. Drunken brawls were commonplace in whatever town was nearby a logging camp, and Coos Bay and North Bend were certainly no different in those days. The bail was a standing $50 per incident, and the men were automatically released Monday morning (probably a bit hung-over). Menasha paid the fine and deducted it from the man's paycheck the next week!

On September 25, 1962, the Tioga Logging Camp bunkhouse burned to the ground. It was never rebuilt; after the fire all the logging crews came from town. Menasha had limited insurance for the personal contents of the bunkhouse, but they made good on every reported lost item—including a pair of false teeth!

In 1999 an evaluation was made of the results of the Irwin and Lyons purchase. From 1954 through 1998 a total of $102 million in profits were made from the sale of timber from these lands. A total of $2 million was expended reforesting the cut-over lands, and the estimated value of the 21,667 acres in 1999 was $71 million. The cash flow from this purchase yielded a compounded rate of return over forty-four years of 13.03 percent—as Mowry Smith Jr. would say, "one of the best investments Menasha ever made."

The other major timberland purchase made during this period was the Al Peirce Lumber Company in 1956.

A total of 11,948 acres was acquired for a price of $1,920,000. The purchase was on contract: $500,000 down at the time of signing the contract in June 1956, $500,000 on the contract anniversary in 1957 and 1958, and the balance due in 1959. Since Peirce had an operating sawmill on Isthmus Slough, the transaction also included the sale of 100 million board feet of logs back to him from all Menasha operations at a set price of $50 per MBF.

Al Peirce Company President Jim Whitty later recalled, "The Menasha agreement also included all the logging equipment we had at the time. The one mistake I made was allowing Junior Arnot to go with the deal. I presumed he would be a lot better off and, that as soon as Menasha got hold of him, they would be smart enough to put him in charge of all their logging operations including Coos River. This they did not do, and after about a year, he purchased a 'cat' from Menasha and went on his own, gyppoing." The Peirce land mostly lay immediately south of Coos Bay where many of the early coal mining operations had been conducted.

Al Peirce was killed in an automobile accident in May 1958, when his car crashed into a redwood tree as he tried to avoid a collision with a logging truck.

During the 1970s the author spent a weekend with a Menasha logging engineer blasting closed the old coal-mine air shafts with dynamite. As we logged the Peirce lands where the old coal mines were located, we found the shafts and air vents still open but surrounded by split-cedar fences. As the loggers progressed through the area, we had the timber fallers mark each of these holes (measuring six to ten feet square) so that logging tractors or an unsuspecting choker-setter would not fall into one. It was great fun as we would lower a case of dynamite into the air vents and set them off. It made a lot of noise, and caused enough concussion that the sides of the shafts caved in and eliminated the treacherous condition.

As we followed our logging with slash burning and tree planting, I received a call on the company radio one July day to inform me that we had a fire in a new plantation and it appeared to be coming out of a coal shaft. Well, it was. Apparently, the fire from the slash burn had found its way into an old coal shaft and burned underground all winter, escaping to the air after the ground dried during the early summer.

Menasha made one additional small timberland acquisition in 1956 when it purchased the Ball Lumber Company located eight miles north of Coos Bay in the community of Hauser. A total of 1,882 acres were purchased for $450,000. Along with the timberland came an old sawmill at Hauser (which burned down in 1957) and a ranch house with racehorse stables and an exercise track for training horses.

The Ball Lumber Company timberlands lands were adjacent to existing Menasha property and contained a reported volume of 16,000,000 board feet. Menasha records show how they booked the purchase price of $450,000 in 1956:

Timber	$290,000
Land	18,250
Logs in inventory	28,700
Advances on logs	12,461
Sawmill	40,000
Logging Equipment	25,000
Residence (above)	30,000
Miscellaneous	5,019
Total	$450,000

Between 1950 and 1969, these three purchases were the only timberland acquisitions made by Menasha in the West.

A total of 377,679,000 board feet of timber was sold from the Menasha holdings in Oregon and Washington from 1951 through 1969—enough timber to build 35,000 single family homes. The total income derived from these sales was $12,715,000. During the same nineteen years 5,695 acres of timberland was sold by Menasha in the West for $1,858,000. Again their investments paid off handsomely!

By 1969, the Menasha inventory of land and timber in Oregon and Washington was as follows:

	ACRES	TIMBER VOLUME
Oregon	73,874	412,310 MBF
Washington	11,097	23,083 MBF
Total	84,971	435,393 MBF

Al Peirce

Jim Whitty, president of the Al Peirce Lumber Company

The Al Peirce Lumber Company mill, shown here in the 1950s, was located on Isthmus Slough at the junction of Highway 101 and the road to Fairview, Oregon (Coos County Historical Museum)

The Ball Lumber Company mill can be seen in the upper center portion of the photo

The "River of Trees" as shown in a Life Magazine article. These are all Menasha logs behind the upper dam on the South Fork Coos River awaiting release down stream to the Menasha plywood mill at tidewater on Coos Bay; approximately four million board feet are visible (Courtesy Brown's Studio, North Bend, Oregon)

Log jams were common during river drives on the Coos River. Turns in the river course, boulders, and stubborn "jack straws" of logs were inevitable (Courtesy of Dow Beckham)

River barge used to clear the obstructions in the Coos River in the early 1900s (Coos County Historical Museum)

The burning of the Upper Dam on the South Fork of the Coos River, December 1957 (Courtesy Coos Bay World)

The Coos River Boom Company

When Menasha purchased the Irwin and Lyons Company in 1954, the company also acquired the Coos River Boom Company—a subsidiary of the Irwin and Lyons Company. Immediately after Menasha took over the operations of the Boom Company, it became caught-up in the controversies surrounding the splashing of logs in Oregon rivers.

The Coos River Boom Company was incorporated on May 18, 1936, for the purpose of "splashing" logs in the South Fork of the Coos River and to build splash dams for holding logs. However, it would take several years before it actually started using the river to store and transport logs. The original subscribers of the Coos River Boom Company were:

Howard Irwin	44 shares
James A. Lyons	43
Ted Pattersen	100
Jane S. Lyons	50
George E. Dix	93
Louise Irwin	50
J. E. Peggs	50
W. H. Woods	50

As time passed, Jim Lyons continued to gain greater control of the Coos River Boom Company and became its president.

Splash dams were important for the timber industry. The development of the sawmill and ship-building industries on Coos Bay in the early 1850s was accompanied by harvesting of the most easily accessible forests on the low hills surrounding the Bay. By the early 1880s, the timber stands close to the mills at tidewater were gone and logging activities shifted further up the Coos River and eventually into the South Fork of the Coos River watershed. Transporting the logs from the "hinterlands" became a real challenge in maintaining an adequate log flow to the mills at tidewater.

The Millicoma, Coquille, Umpqua, Siuslaw, and Coos Rivers all became "logging roads" for transporting logs from the woods to the mill. As the logs settled into the gentle tidal reaches of Coos Bay, they would be collected by small boats, assembled into "log rafts," and stored in "log booms" for later towing to the mills. Menasha maintained the Waterford Boom and the Cooston Boom on the east side of the Bay, where old pilings can still be seen poking out of the mud flats at low tide.

The logs that arrived at company booms on the Bay were without any identifying marks as to whose they were. There was no way of telling who owned the logs that floated by the boom boats, or broke loose from their tie-ups along the river banks during heavy winter freshets and ended up on the tide flats around the Bay. A "gentleman's agreement" between the companies who used the Coos River system to transport their logs prevented arguments from ever developing among them.

However an interesting set of characters evolved to help themselves to Menasha (and other timber companies') logs stored in these log booms. These were the notorious Coos Bay log pirates.

INCORPORATED UNDER THE LAWS OF THE STATE OF OREGON

NUMBER 6 — SHARES 1

Coos River Boom Company

Marshfield, Oregon

This Certifies that J. L. Hawkins is the owner of One Shares of the Capital Stock of COOS RIVER BOOM COMPANY transferable only on the Books of the Corporation by the holder hereof in person or by duly authorized Attorney on surrender of this Certificate properly endorsed.

In Witness Whereof the duly authorized officers of this Corporation have hereunto subscribed their names and caused the corporate seal to be hereto affixed at Marshfield, Oregon this 24th day of February A.D. 1942

Shares $100 Each

INCORPORATED UNDER THE LAWS OF OREGON

Coos River Boom Company

Coos Bay, Oregon

This Certifies that JAMES A. LYONS is the owner of TWENTY EIGHT and Five/Eighths Shares of the Capital Stock of COOS RIVER BOOM COMPANY transferable only on the books of the Corporation by the holder hereof in person or by Attorney upon surrender of this Certificate properly endorsed.

IN WITNESS WHEREOF the said Corporation has caused this Certificate to be signed by its duly authorized officers and its Corporate Seal to be hereunto affixed this 15th day of NOVEMBER A.D. 19 47

SECRETARY — PRESIDENT

SHARES $100.00 EACH

It took several days for logs released from the Upper Dam to reach the Bay where they would be assembled in rafts and tied to pilings across from the mill (Courtesy of Warren Dick)

The Upper Dam, twenty-one miles above Dellwood, with the lake at the "full discharge level," South Fork Coos River, 1956 (Courtesy of Warren Dick)

The Upper Dam on the South Fork of the Coos River releasing its logs, spring 1956 (Courtesy of Warren Dick)

They often employed ingenious techniques to extract logs from the tightly corralled logs in the boom or "found abandoned" upon a nearby beach. One such approach involved lashing a boom log (a single row of logs attached to one another by a cable end-to-end that surrounded each raft of logs as they lay among the piling) in the dead of night to an upright piling during high tide, then waiting for the tide to lower. With the boom log now suspended above the water, they would enter the log rafts, extract two or three logs, and tow them to the tide flats for later removal. Since there was no way of identifying the logs, they became what is known legally as derelict logs available for taking by anyone with the ability to retrieve them. These same logs might then be sold to the original owner without them knowing that they were theirs in the first place—at least so the log pirates argued!

A game of cat and mouse finally evolved as the timber owners banded together to form the Coos Bay Log Patrol Association. They hired a security guard to patrol the Bay to collect derelict logs and catch the log pirates red-handed.

The use of water to transport logs was not something invented in the Pacific Northwest. Undoubtedly, the early pioneers in New England used their rivers to transport wood from one spot to another. Menasha records also show that C. R. Smith used Midwestern rivers and lakes for storage and transport of logs to support their sawmills and wooden ware business in Wisconsin.

However, things always did not go as planned on every log drive.

Snags, boulders, and other obstructions often blocked the river course used to float logs down to the mills. These obstructions had to be removed in order to make the watercourse predictable and safe. One method used to clear the deeper channel of snags and sunken logs was a river barge.

In the upper reaches of the river basins where the water was too shallow for a river barge to work, the method of choice for removing obstructions was dynamite. Dow Beckham, one-time superintendent for the Coos River Boom Company, described such a procedure: "There were lots of obstructions in the upper portion of the Coos River—mainly boulders. On one occasion we packed three cases of dynamite under a big rock in the center of the river, connected a sufficient length of fuse to allow time to get away from the blast, and then ran like hell for a nearby old growth tree behind which to hide and hope the rain of the exploding rocks would not ricochet off an adjacent tree."

Since the major rivers running into Coos Bay do not originate from a lake (as they often do in the Midwest), logs could only be moved in the rivers from the upper parts of the watershed during winter freshets. The river was simply too shallow to float logs during the summer months.

In order to supply a more predictable source of logs to the mills in Coos Bay, artificial ponds were created in the rivers. These ponds provided two functions: to provide an inventory storage system for holding logs and to provide a surge of water to help propel the logs downstream when the dam was breached. Wooden dams, called splash dams, were built on the Millicoma, Coos, and Coquille Rivers for this purpose.

Dow Beckham remembered how difficult dam construction on the South Fork of the Coos River could be: "Supplies—heavy spools of cable, boxes of bolts, sacks of cement, dynamite, fuel, and food for the men—had to be hauled over three miles of precipitous 'cat' roads from the end of the plank road which extended over Cathcart Mountain to the south side of the river near

Dellwood." The author recalls driving on that same plank road thirty years later. The planks would get dislodged by the tires and get wedged in the frame of his Menasha pickup!

During the winter freshets of 1941 the Lower Dam was washed out and a new one was built about three miles farther upstream. In April 1942, road builders working on a nearby road set off a dynamite blast that knocked out a brace behind the dam causing that dam to fail, so a third "Lower Dam" was built at the same spot later that summer.

Logs would be dumped into the pond behind the dam and aligned parallel with the channel to wait their turn to pass through the dam opening once the river-run began. Often millions of board feet would be stored in these makeshift lakes on the Coos.

Just before the gates to the splash dam were released, a Menasha employee would drive down the Coos River road below the dam with a bullhorn telling fishermen or swimmers in the river that a river run was about to begin.

Once the dam was "breached" there was no turning back; and the log drive was on!

The second dam, known as the Tioga, or Upper Dam, was completed in 1945. It was the largest and last operating splash dam in the Pacific Northwest. Its spillway was fifty-two feet above the river surface. It was about twelve miles farther upstream than the lower dam. This dam enabled water to back up about one and a half miles into the Williams River and was the subject of a *Life* magazine photo called "River of Trees."

The top part of the dam extended about 150 feet across the river and was about 30 feet high. It had two gates: one fifty-five feet wide and eighteen feet high, the other twenty-four feet high and twenty-four feet wide. The gates were for controlling the water, thus making it possible for logs to pass through them.

The franchise for splashing logs in Oregon was an emergency measure for obtaining logs during the war years of 1940 to 1946, as there were no usable roads into the backcountry. However, the Oregon Fish and Game Commission and local sport fishing groups vehemently opposed the practice of splashing logs in the river. William Vaughn, owner of the Coos Bay Logging Company, would testify against splashing logs on the Coos whenever the opportunity arose.

The remnants of the Lower Dam on the South Fork of the Coos River can only be seen at low water during the summer months. The riparian areas adjacent to the river show no evidence of the damage caused by the running of logs in the river and clearly demonstrate the resiliency of nature

Dow Beckham remembers the arguments that regularly occurred between Vaughn and Jim Lyons. Vaughn had built a truck road on the north side of the South Fork Coos River from Dellwood easterly to what is now known as the 7-mile bridge. Vaughn wanted the logging and log-hauling business out of the drainage and Lyons wanted to transport his logs via water to his mill at tidewater.

In October 1945, the Portland office of the United States War Department called a meeting to air grievances about the practices of the Coos River Boom Company and the problem of loose logs in the upper tidal sections of South Coos River, but it was not until April 1946 that the Secretary of War laid down regulations affecting log splashing on the South Coos River. As strange as it seems in hindsight, the government would allow the "running of logs" down the Coos River only on Mondays and only during the months of August, September, and October. During the months November through July, logs could only be floated downriver on Tuesdays.

All that is left of the Upper Dam on the South Fork of the Coos River is the foundation and a remnant of the old gate buried in the silt

As Dow Beckham recalls: "The rules were simply impossible to follow except on rare occasions. We soon violated the rules in order to continue with our work. We were never challenged about

Long log truck on way to mill (Oregon Historical Society 16864)

Road building required timber fallers to cut the standing trees on the right-of-way so a tractor could pull the logs out of the way of the actual road construction work. Note how far the "choker setters" had to pull the one and one quarter-inch steel "bull line" to reach the turn of logs

Building a road on Menasha timberlands in Oregon (ca. 1960) (Courtesy of Warren Dick)

ROAD BUILDING

Concurrent with the use of splash dams on the Coos River came the construction of truck roads to access the timber growing higher on the hillsides. The logs hauled by trucks on these precariously steep and narrow logging roads would either be delivered to the holding ponds behind the splash dams or delivered directly to tidewater to be assembled into rafts for later towing to the mill. Logs for the mills that were not located on the Bay were inventoried in large, flat storage yards.

During the early part of this period, little formal engineering was done prior to the actual road construction. Often the task of "laying out the road" fell to the logging superintendent whose main focus was to get the road to the next logging site as quickly and cheaply as possible. They knew the capacities of their road construction equipment and of the log trucks that would pull their heavy loads of logs away from the landing and down the winding roadway.

Road building on the steep side-slopes of the Coast Range is a dangerous proposition. Inexperienced road builders might push insufficient dirt ahead of them on the slopes and end up with a treacherous, if not fatal, ride to the bottom of the canyon. "Cat skinners" (men who operated tractors in a logging/road construction operation) were a breed all their own.

our flagrant abuse of these rules, even when the steelheaders were pushing so hard to eliminate the practice."

The opposition to log splashing on the Coos grew to a point that something needed to be done to find a compromise between those needing the river to transport logs and those wanting the river free from logs. Menasha stepped into this fracas when it purchased the Irwin and Lyons Company, including the Coos River Boom Company, in 1954.

On February 16, 1956, a local meeting was held that drew a vocal group of ardent opponents to log splashing on the Coos River. Al Stinchfield, manager of the Menasha Plywood Corporation and president of the Coos River Boom Company, represented Menasha. The sole purpose of the meeting was to determine steps leading to the termination of log splashing on the Coos River.

On October 9, 1956, the Oregon legislature banned the use of rivers in Oregon for running logs and the Coos River Boom Company was given until June 30, 1957, to remove all logs from the river and to remove both dams. In the last week of December 1957 Menasha set fire to both splash dams and thus ended almost two decades of log running on the Coos.

Even though log splashing was no longer permitted in the upper reaches of the rivers, the lower part of the Coos River was still available for transporting logs to the mills at tidewater. The Weyerhaeuser Company continued to use the lower portions of the rivers for the movement of logs from their staging areas at Dellwood on the South Fork of the Coos River and at Allegany on the Millicoma River. In 1991, Weyerhaeuser stopped using both rivers for the transport of logs. As shown in photos of Coos Bay in earlier chapters, the Bay, rivers, and side sloughs were all used extensively for storing logs in "booms" awaiting delivery to a mill or to shipside for export to Asia.

After the Menasha splash dams were eliminated from the upper portion of the Coos River, the "off-highway rigs" would drop their cargo of logs at the Menasha "dump" just above Dellwood. After the logs were in the water, they would be pushed into rafts by small boats called "log broncs" for later towing to the mill (Courtesy of Warren Dick)

Tug towing two flat-rafts of logs on the Coos River below Dellwood, 1962 (University of Oregon)

Road grades in the South Coos River drainage were often too steep for a truck to carry heavy loads uphill unassisted. Here, a log truck helps pull a "low-boy" hauling a Menasha yarder up the "rock side" of the Mink Creek road on the Oregon tree farm. This type of yarder weighed in excess of 185,000 pounds. Note the log that fell off a truck earlier that day. On occasion, a bulldozer would help push or pull a heavy-laden truck up the very steep roads (Courtesy of Warren Dick)

An off-highway Menasha log truck starts the steep downgrade to the Mink Creek bridge crossing the South Fork Coos River. Note the brow-log buried in the road fill in order to make the road wide enough for the trucks to make the turn. Cables were wrapped around the outside log(s) and anchored to "dead-men," which in turn were anchored into rock or a large log buried in the roadbed beneath the truck (Courtesy of Warren Dick)

Road graders were used to spread gravel on the dirt road surfaces and to keep the running surface of the main roads free of potholes that damaged the tires of the log trucks (Courtesy of Warren Dick)

During the operation of the Menasha plywood mill, extensive logging of both Menasha and government lands was required to meet the demand of the mill. In order to reduce the overall costs of logs delivered to the mill, larger and larger logging trucks were introduced to haul the logs on Menasha private roads. These "off-highway" logging trucks had twelve-foot wide bunks (stakes on the outside of the log load) in contrast to the eight-foot bunks allowed on the public highways. A fully loaded off-highway truck would weigh in excess of 130,000 pounds compared to the legal highway load of 80,000 pounds

Small streams were crossed using culverts ranging from twelve to forty-eight inches in diameter, but the water flow at Shotgun Creek on the Menasha tree farm in Coos County required a much bigger culvert called a "multi-plate." Menasha employees (left to right) Ron Durham Sr., Frank Lyon, and Ed Wade work on the installation

Introduced in the 1970s by Weyerhaeuser was the "triple" off-highway truck. These behemoths could haul 18,000–20,000 board feet compared to the normal highway log truckload of 4,500 board feet

Outcroppings of rock in the roadway called for a special piece of equipment called a track drill to remove the obstacle. First several holes would be drilled into the rock face by the air-driven track drill, the holes would be packed with explosives and the rock would be shattered. A large tractor would then push the rocks to the side and continue to excavate the dirt as the road progressed to the next logging site

Bridges had to be built to support the massive weight of these fully loaded trucks and giant mobile yarders. The construction of the wooden bridge at the junction of Mink Creek and the South Fork of the Coos River in the 1950s shows how local loggers and forest engineers utilized available resources to span the river crossings. Here the bridge is supporting the Menasha Logging off-highway truck # L-21 with twelve-foot bunks. The total weight of the truck and its logs topped 130,000 pounds (Courtesy of Warren Dick)

Blasting a rock outcrop out of the way of the road construction

Logging is a very dangerous occupation. Bill Haresnape, the Menasha personnel manager, recalls dealing with at least six tragic deaths in the woods during his tour with Menasha (Courtesy of Warren Dick)

The use of culverts to divert water from eroding the roadbed is not only required by the laws administered by the Oregon and Washington Departments of Forestry, they are necessary to keep the running surface smooth

Norm Webb topping a spar tree on Menasha timberland, 1961 (Courtesy of Warren Dick)

chapter seven

Cruisers, Fallers, and High-Climbers

Until 1950, there were no college graduates with formal degrees in forestry working for Menasha. Then in 1950 Ted Ellingsen, who grew up in Coos and Curry Counties and graduated in forestry from Oregon State University, took a job with the company as a timber cruiser.

At the time Ellingsen came to work, Ned Elder had been Menasha's timber cruiser for years and Norm Laird was his compass man. Elder was one of the old-time timber cruisers who built their skills over time without any formal training in statistics or forest sampling. He helped the young Ellingsen ease into the spot as head cruiser for Menasha. Elder died in 1954.

It was the timber cruiser's job to estimate the amount of recoverable volume in each tree as well as to determine the total volume for a given tract of timberland. As the cruiser studied each tree, his compass man would keep the project progressing on a particular "heading" and pace off the distance for each "tally." The estimates by the timber cruisers were critical to the scheduling the company's plywood mill operation in North Bend.

Ted was not a newcomer to the woods of Coos County. Having grown up in the area, he worked as a cruiser-assistant for the county during summers while going to forestry school. Asked to describe the "old-time cruisers" Ellingsen said, "They were like a secret society. They were always closed-mouthed and would not talk about their cruising methods or procedures. The simple reason they did not discuss their procedures was because they did not have any [standard methods or procedures]!"

In defense of those early timber cruisers, there were really no tables that accurately computed the volume of Douglas-fir timber until Mason, Bruce, and Girard first published their cruising tables in 1947. As one story goes, some cruisers simply memorized the recoverable lumber volume in a forty foot log of each diameter above eighteen inches.

In those early days if the cruiser's information was questioned, it may have been due to the fact that the loggers who came in behind the cruiser to harvest the timber would take only those trees and logs they thought the mill wanted or that they could harvest easily, leaving the rest of the trees and wood behind. It was a constantly moving target for the early foresters to estimate the amount of recoverable volume from a tract of timber so that the mills could plan their production schedules. The loggers always thought they knew better than some college-trained forester!

Ellingsen remembered cruising the Irwin and Lyons property before it was purchased for Menasha by Mowry Smith Sr. Apparently Menasha and the manager of Irwin and Lyons had agreed to independently cruise the property and then have the cruisers reconcile their differences on each tract.

Each cruiser had their own "brand", a mark left on trees indicating the cruiser had passed by. Ted Ellingsen's mark was Ɛ

Ned Elder, a timber cruiser for Menasha in the 1940s and 1950s

Norm Laird, compass man for Ned Elder and Ted Ellingsen, shown here on a cruising project in Curry County, Oregon

TIMBER CRUISER EQUIPMENT

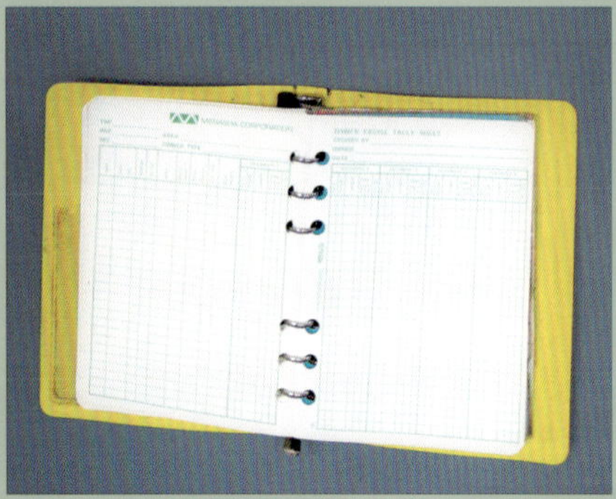

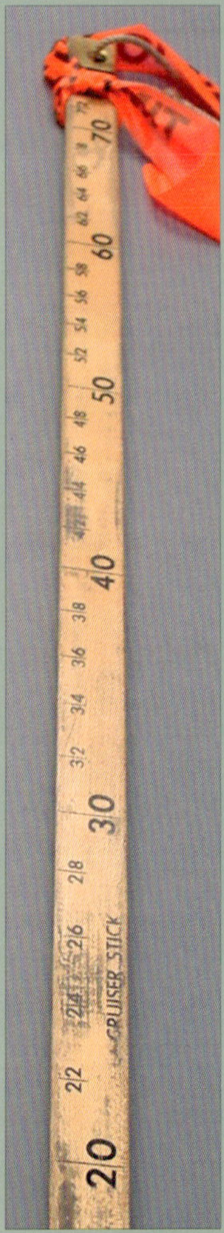

Ellingsen relayed the following story:

> *Warren Dick was the general forester [without formal forestry training] for Irwin and Lyons, and he was assigned the task of cruising for them. As Ted and Norm Laird began their forest sampling they noticed that Warren had hung his strips of plastic ribbon marking his cruising trail along the ridge tops, rather than hiking down to the bottoms of the canyons as Ted and Norm did. Warren must have simply peered over the edge of the steep hillsides and guessed. However, after it was all said and done, the two different cruises were remarkably close.*

The tools of these early foresters were fairly basic, consisting of a good pair of caulked boots for traction on the slippery moss-covered ground, a lightweight rain jacket to keep some rain off, a hand compass for direction, a Biltmore stick for measuring tree diameters, a topographic map for determining location, a notebook with write-in-the-rain paper, an axe to mark the property corners with their unique brand, and of course the infamous red felt hat.

After the cruisers came the loggers—who, as they cut their way east into the rugged Coos County coast mountains, developed new logging techniques. The logs of the Coos River basin were exceptionally large; the big logs, combined with the steep terrain of the basin and heavy equipment, daily put the logger's skills to the test.

It was not long before gas- or diesel-powered donkeys replaced the steam-powered winches of the early twentieth century. This made the donkeys far more mobile.

To move the donkey and its underlying sled from one logging site to another, cables were hooked around a stump in the desired direction of travel and the sled simply winched its way from one stump to the next by means of pulleys. The process was repeated over and over until the donkey was repositioned for the next logging operation.

A diesel-powered donkey yards logs down a ravine (Courtesy of Warren Dick)

Tree fallers (Courtesy of Warren Dick)

The donkey was positioned away from the "spar tree" and provided the power to drag the huge logs up the hill. A second donkey would be positioned to provide power to the heel boom loader. A suspended log can be seen in the very right of the photo. Note the quantity of guy wires attached to the spar tree (Courtesy of Warren Dick)

This jumble of logs at the bottom of a ravine in the South Coos River drainage resulted from the timber being "swung" from one spar tree to another. Here it is being yarded to the final landing, where the logs will be loaded onto log trucks for the trip to the mill (Courtesy of Warren Dick)

Note the size of the men setting chokers in comparison to the big logs being yarded to the landing (Courtesy of Warren Dick)

The steep terrain of the Coos River basin and the massive sizes of the trees also created challenges to the loggers who might drag or "swing" logs as much as a mile before they were dumped into the river for their final journey by water to the mills.

As access roads were developed to reach deeper into the Coos County forests, ingenious cable systems were developed that were designed to drag logs uphill to a central "landing" site where the logs would then be sorted and loaded on trucks for transport downhill to the mills. This process involved utilizing a tall, strong Douglas-fir tree growing in just the right location. Cables and pulleys would be attached high in the tree in order to create additional elevation over the traditional ground-skidding technique so that "drag" would be reduced as the logs were hauled in. Logging engineers refer to this mechanical advantage as "deflection."

A "high-climber" would scamper up the selected tree using nothing more that a loose rope around the stem of the tree; he had two-inch spikes strapped to the inner sides of his boots to penetrate the bark for traction; an axe and a crosscut saw were hooked to his belt.

The high-climber tied a small (usually three-eighths-inch hemp) rope to his belt. When he had topped the tree, he used the rope to pull up a small fifteen-pound pass block (pulley) that he fastened near the top of the spar pole. With the use of the same rope, he then threaded a three-eighths-inch wire cable coming from the donkey spools through the pass block back down to the ground where tree plates, guy lines, bull block, haul-back block, and other equipment were attached to the cable so that the donkey could lift these to the top of the spar pole.

For loading logs onto trucks a "hayrack" or heel-loading boom would be attached part way up the spar pole. At this point the high climber's job was done.

Once the tree was fully rigged and all the timber on a particular "setting" had been felled, the actual logging operation would begin.

Later, as diesel-powered machinery became adapted to the logging operation, mobile yarders replaced the donkey sleds, and spar trees and grapple loaders replaced the heel booms attached to the spar trees.

Menasha timber fallers Red Weybright and Al Nighswonger "dump one down the hill" (Courtesy of Warren Dick)

Gasoline and diesel-powered equipment started replacing steam in the 1930s; by the 1950s, when this photo was taken, the internal combustion engine provided the power to skid logs (Pacific County Historical Society)

Moving these large "donkeys" up the steep side slopes was extremely dangerous (Courtesy of Warren Dick)

"Plumbing" an old-growth Douglas-fir on Menasha lands. This tree was probably more than 200 feet tall (Courtesy of Warren Dick)

Timber fallers, Al Nighswonger (left) and "Red" Weybright (right) (Courtesy of Warren Dick)

FALLING AND YARDING

Timber fallers "face" an old-growth Douglas-fir (Courtesy of Warren Dick)

Menasha higher-climber Norm Webb readies the "spar tree" by first removing the limbs on his way up and cutting off the top of the tree in preparation for rigging it with pulleys. On occasion, the high-climber, after topping the tree, would actually climb on top of the "stump" to have a cigarette and enjoy the view from atop the 120-foot high pole! (Courtesy of Warren Dick)

High-climber with his cut-off saw hooked to his belt making his way up a "spar tree" (Courtesy of Warren Dick)

Menasha employees Norm Webb and Paul Coleman rig the spar tree with pulleys (Courtesy of Warren Dick)

This Menasha spar tree is now fully rigged for yarding. The two horizontal logs attached to the spar tree, called a heel boom, were used to load logs onto awaiting trucks, or move logs away from the spar tree (Courtesy of Warren Dick)

A heel boom attached to the spar tree is seen here loading a Sterling log truck on Menasha lands in the late 1940s (Courtesy of Warren Dick)

The Menasha D-6 tractor pulling the Rome disk across the grassy fields south of Coos Bay in the summer of 1957. This was a new process tried by Menasha foresters to convert abandoned pastureland to a healthy forest

chapter eight

The Value of Good Forestry

Current laws governing forest operations in Oregon date to 1941 when the government addressed the reforestation of timberland following a logging operation. At the same time, the American Tree Farm system came into existence as a way of promoting good forestry practices throughout the United States, primarily on large industrial timberland ownerships.

As the profession of forestry began to evolve, new ideas and practices—usually mimicking nature's processes—were tried in the West. In the early 1950s, it was common to collect Douglas-fir seed from trees growing in the local area during times when cone production was particularly abundant and then keep the seed in cold storage until needed, often five to seven years into the future. Foresters found that Douglas-fir trees produce a good seed crop about every seven years and the seed collected during these bumper years could be stored in freezers for up to a decade without significant deterioration.

Three facts stand out and are often overlooked in the debate over clearcutting as a forestry practice in the West. First, Douglas-fir seedlings cannot survive in the shade. Once brush or trees of other species overtop the young seedlings, they cannot survive. Second, it is practically impossible to harvest timber on the steep side slopes of the Oregon Coast Range without clearcutting. And finally, the clearcut practice mimics the way nature created the massive Douglas-fir stands in the first place, through large indiscriminate wildfires or fires started by Native Americans in order to manage their food sources. Foresters simply replicated what nature had evolved.

Because much of the timber harvested by Menasha during the 1950s originated from the catastrophic fire of 1765, many of the trees were "over-ripe" and contained excessive amounts of unusable wood. This rotten wood, left in the woods after cutting along with limbs and treetops, was collectively called "slash."

Logging old-growth timber during this period often created large concentrations of slash, rotten logs, and debris in the bottom of the steep draws. To reduce future fire hazards on the logging site, the slash had to be burned via controlled fires.

The author recalls, as a young forester for Menasha, the excitement of using old used truck tires in the slash burning efforts. The cavity of the tire would be filled with a mixture of diesel oil and gasoline, ignited, and rolled off the hillside in hopes that the bouncing tire would splash the fuel and spread the flames as it traveled to the bottom of the canyon. Often this was more fun than practical as slash burning in those days was usually done in the middle of winter when such antics were mostly useless.

Usually the forester would have to climb to the bottom of the canyon to get the slash to burn. Here is where real fire-building skills came to bear, as it would be during late November or

Boyd Swenson, chief forester for Menasha in the 1950s, shows the results of the reforestation effort on company lands near Bandon, 1958

December when rain insured that no respectable fire would try to run away. The best materials for lighting a fire in the wet slash were a gallon jug of diesel oil in an old Purex bleach bottle, a box of wooden matches, and a source of "squaw-wood" (stump slivers containing heavy amounts of pitch). The technique was to start a small fire and add pieces of ever-increasing size until the fire was ready to "run on its own." Nothing can describe the eerie sounds as the fire in the canyon built momentum. However, seldom did a controlled burn go beyond its boundary.

Hiller 12-E contract helicopter conducting an aerial seeding operation on Menasha lands in Oregon

After the brush had been removed by a controlled burn and there were no "hot spots" remaining on the site, foresters were ready to begin the reforestation process. They commonly used small helicopters equipped with a spinning spreader to disperse Douglas-fir seed evenly over the prepared site. This method, called "aerial seeding," sprinkled about 30,000 seeds over each clear-cut acre.

As mentioned earlier, Douglas-fir, the primary species of forests in the Pacific Northwest, produces a bumper crop of seed about every seven years on average. Because of this seed cycle, foresters had to plan years in advance in anticipation of the amount of seed needed in the aerial seeding process. Orders would go out to cone-buying stations within the region where future reforestation efforts were anticipated. Professional "cone pickers" would climb trees growing in the nearby forest and pick cones from the upper branches. The less reputable pickers might strip the trees of their branches and pick the cones from the ground, or even cut the smaller trees down to obtain their bounty. The price of cones was around $10 per bushel and a bushel produced about a pound of seed, enough to seed a newly logged area.

Mowry Smith Sr. plays with Menasha's new HD-21 tractor as it sits awaiting a canopy in the Menasha Plywood warehouse in North Bend. This machine would be used to log the scattered trees on the recently acquired Al Peirce lands and clear the brush ahead of the disking project. However, the main work of the tractor was to build roads on company lands in Oregon

Another common source for cones was family outings, usually only lasting a short Saturday as the pitch accumulating on the hands, between the fingers, and on the children's clothing was soon judged not worth the $10 rate—that is, unless a wandering youngster happened upon a cache of cones stashed away by a hard-working squirrel.

In those early days when professional forestry was in its infancy in the West, quality control of the seed source was poor at best. The hope was that the cones actually came from the forest nearby the Menasha tree farms. Unscrupulous cone buyers might sell the same bushel of cones to an unsuspecting forest seed company two and three times!

The practice of aerial seeding was often a hit-or-miss proposition. If the winter conditions were not just right or the following summer happened to be unusually hot and dry, the seedlings sprouting from the seed would often perish. Consequently, Menasha foresters began hand-planting seedlings in the early 1950s as a more predictable reforestation method, and the practice of professional forestry really began on Menasha lands at that time.

A new Menasha fire truck is inspected by (l. to r.) Tom Stinchfield, Mary Smith (Mrs. Mowry Smith Jr.), Mowry Smith III (Trip), and Mowry Smith Jr. (ca. 1950)

Some of the land in the Al Peirce acquisition in 1956 contained brush-choked hillsides of an abandoned sheep pasture. This required some new and creative ideas in order to get the land back into forest production. As the Coos Bay *World* newspaper reported on December 12, 1957:

> *A North Bend Plywood manufacturer has become one of the first companies in Oregon to employ large-scale site preparation techniques in reforesting some 6,000 acres of barren land.*
>
> *Menasha Wooden Ware Corporation is seeding and hand planting trees on land it recently purchased just southwest of the city where natural seed sources are absent.*

In undertaking the vast reforestation project, the firm has become one of the first in the Northwest to utilize a disk plowing harrow on a broadcast seeding site and is one of the first in Oregon to scarify—a forester's term for "scratching"—forest land with a bulldozer. Ralph Horn, Timber Manager, reported that the equipment is a D-6 caterpillar pulling the two-gang disks. The other machine used is an Allis-Chalmers HD-21 equipped with a brush blade to clear any brush out of the way. Boyd Swenson, Chief Forester of the company is in charge of the project. . . .

But since it takes scores of years to grow a good peeler log for plywood use, why is the company—the parent firm of Menasha Plywood Corporation—spending money on reforestation? "We don't want to see good timber-growing land stand idle," explains Al Stinchfield, company general manager. "Nobody knows what the plywood situation will be twenty or thirty years from now, but we're certain of one thing—there'll always be a market for trees."

From 1950 to 1955, the forestry office of Menasha was in the American Bank building in Coos Bay. The forestry operation moved into a small office at the Menasha Plywood building from 1955 to 1962, but the forest engineering employees worked out of the back office at Mill B during this same time. From 1962 to 1968, the company forestry and engineering operations were conducted out of the upstairs of the old Citizen's Bank building at the corner of Highway 101 North and Virginia Street.

When the Menasha Plywood operation closed in 1967 and no longer needed the office building it had built four years earlier, the forestry office moved into the building. This is the same location where the current Menasha Forest Products Corporation headquarters is located.

The forestry office of Menasha was located on the second floor of the Citizen's Bank Building in North Bend. The offices were in the corner overlooking the Bay. Robert Banks of Kruse and Banks Shipbuilding Company owned the building, which still (as of 2005) stands at the corner of Virginia Street and Highway 101 North (Oregon Historical Society 98120)

Menasha was recognized for its good forestry work in the late 1950s by qualifying as a new member of the American Tree Farm System (left to right) Bill Hagenstein, executive director of the Industrial Forestry Association; Menasha representatives Ralph Horn, Al Stinchfield, and Mowry Smith Jr.; and Calvin L. Smith of Medford, Oregon, manager of the IFA

The Columbus Day Storm

At 10:10 a.m. on the morning of October 12, 1962, the following advisory went out by teletype to all points: "A very deep vigorous storm that is the remnant of a Pacific Ocean tropical storm is moving in a northeasterly direction and is expected to be centered over the ocean near the mouth of the Columbia River this afternoon. Ship reports near this storm center indicate that it is a very dangerous storm. Whole gale warnings are displayed on the north California coast and gale warnings on the Oregon and Washington coast are expected to be changed to whole gale warnings in the next few hours."

A week later, after the storm wreaked devastation across the state, the Portland (Oregon) *Journal* reported:

Oregon's large timber loss during the recent windstorm may be just the beginning of what could become a major loss of valuable trees from attacks of bark beetles. Oregon State University's forest entomologists said the large amount of blown-down timber left in the wake of the storm would provide breeding location and food supply for beetles that will emerge next March or April. Salvage logging will remove the dead and downed timber and cut the emerging beetles from their food supply. And then it hit the Oregon coast. It was not only a furious storm but also a big one. It was a cyclone, wind blowing in a circle estimated at 500 miles across. Winds were

Menasha was not spared the wrath of Hurricane Frieda, as millions of board feet of old growth timber in the Coquille and South Fork Coos River watersheds were smashed (Courtesy of Warren Dick)

A view of the Charleston, Oregon, boat basin in Coos Bay during the Columbus Day Storm on October 12, 1962 (Hurricane Frieda) (Courtesy of Jerry Bacon)

clocked in Portland Oregon at 116 mph and 152 mph at Hauser just north of Coos Bay. It was called Hurricane Frieda. . . . The Pacific Northwest's big windstorms of 1949 and 1951 unleashed a beetle attack that ruined 3 billion board feet of timber.

The storm caused $235 million in property damage across Oregon and blew down fifteen billion board feet of timber (enough wood to build 1,500,000 homes) valued at $750 million ($4.75 billion in 2004 dollars). The inability of the local mills to harvest and utilize the massive amount of timber blown down by the storm before natural rot set in launched the log export business to Japan. Massive harvesting efforts went forth on all ownerships and a portion of the logs were loaded onboard ships destined for Japanese sawmills.

Two years later nature struck once again, this time at the Oxbow some fifty miles north of Coos Bay, when a fire driven by excessively strong dry winds ravaged 42,000 acres of timberland. However, no Menasha lands were involved.

As noted at the end of Chapter Five, Menasha decided in 1967 to close its plywood and stud mill and its logging operations. However it would continue to operate the Wood Fibre Division under Dick Hansen and keep the Oregon and Washington forestlands intact. The North Bend paperboard mill located on Jordon Cove would continue operation under Menasha control for fourteen more years. This was a major turning point in the history of Menasha's involvement in the timber business in the West.

By the end of this era, Menasha's forestry programs consisted of burning the slash left by the logging operation to reduce the fire danger, followed by aerial seeding of the cut-over lands with Douglas-fir seed, then interplanting seedlings where the aerial seeding effort had failed. For the most part, these efforts to create the next generation of trees on Menasha lands were successful.

A final note on the western forestry practices of Menasha during the 1950s and 1960s: Wildfires were, as they are today, a nemesis of any timber owner, and Menasha invested heavily in its ability to fight wildfires.

Tree planters

A buffer strip protecting a creek on Menasha timberland

Carefully tended nursery seedlings being transplanted to reforest Menasha timberlands

chapter nine

The 1970s

Following the closure and sale of the Menasha plywood/stud mill in North Bend in 1967 and the phasing out of Menasha's logging by 1969, the company was faced with the problem of creating income from timberlands that no longer had a captive log-converting facility. Who would purchase Menasha's timber?

A plan for selling standing timber harvesting rights to specific tracts of Menasha timberland was devised–much the same way Menasha sold its timber fifty years earlier. The firm of Mason, Bruce, and Girard, consulting foresters in Portland, Oregon, was instrumental in assisting Menasha in making this shift.

However, the tree farm itself had a problem that was not readily apparent to the outside world, but which held great concern for the viability of Menasha. The "age-class distribution" of the trees growing on Menasha timberlands was such that management could predict a shortage of mature timber to harvest by the mid-1980s, causing a gap in future earnings that was likely to last for more than a decade. Something had to be done.

Because the timber assets of Menasha had been a major source of profits for decades, company President and CEO at the time Richard L. "Dick" Johnson quickly began searching for a solution to the dilemma. In 1969, Carl Newport and Lou Alexander, senior partners in the Mason, Bruce, and Girard, were brought in to assess the problem and suggest a solution. The author was just finishing graduate school at Yale at this time and was brought on board by Mowry Smith Jr. to learn from the consultants.

Months of inventory and analysis confirmed the gap in the ages of Menasha's timber. The only viable solution was to slowly decelerate the liquidation of the remaining mature timber to the point where it could be used to "fill the gap" in income ten to fifteen years in the future. A sustained-yield harvesting level was established to meet that objective.

Fortunately, timber prices literally "shot through the ceiling" during the 1970s and Menasha timber operations became highly prized by company accountants, generating over 25 percent of the total company profits for the decade. Menasha was changing, expanding its other businesses in plastics, paper, and paper converting, along with becoming actively involved in other evolving industries. The cash generated by the timber division helped to fuel this company-wide growth.

For multiple tax reasons, Menasha sold its timber during this era based upon a "retention of economic interest"–a technical term that required the purchaser to pay for only the timber they removed from the logging site. Species prices were predetermined through a sealed bidding process and buyers were given a six-month time frame within which to harvest and remove the timber. Menasha did very little actual logging themselves, unlike the period when its timberlands were integrated with company logging operations and the plywood mill during the prior 20-year period.

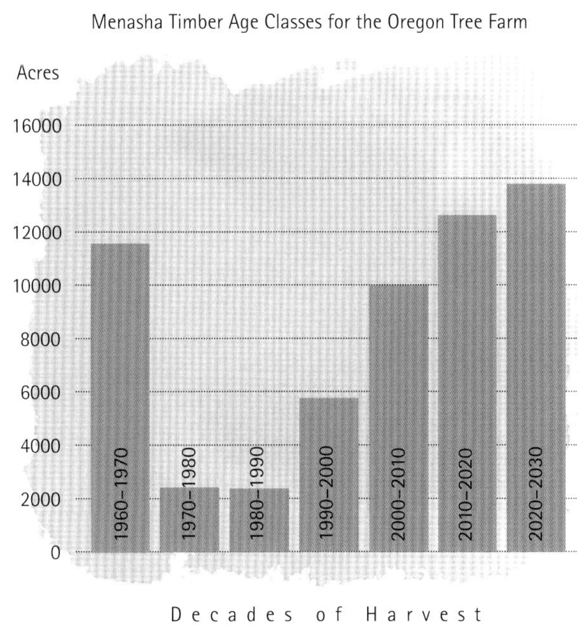

The age gap in Menasha's forest inventory, estimated in 1969, can be seen in the center of the chart above

Timber fallers using a Dellwood Timber Tipper to fall big trees uphill on a Menasha timber sale (ca. 1976)

The process of selling Menasha timber was quite simple: The company auctioned small tracts of timber to local mills through a sealed bidding process. Only the timber was sold; the land remained Menasha's. Even though the company no longer owned any forest products manufacturing plant, there still remained a need for predictable results from the sales, so Menasha timber cruisers would estimate the recoverable volume on the advertised stumpage sale long in advance of the actual auction, just as they had done for the Menasha Plywood mill in the 1950s and 1960s. Menasha foresters often made wagers as to whose field cruise would best approximate the final harvest volume.

A detailed timber sale prospectus was then prepared and mailed to prospective purchasers. At the appointed hour of the auction, there were often as many as a dozen local mill owners and loggers lined up at the Menasha office to submit their best offer. Profits came easily. But Menasha's President Johnson remained concerned about the upcoming age-class gap.

The first timber-cutting contract (stumpage sale) of the 1970s was sold to Leep Logging Company of Myrtle Point, Oregon, and yielded \$313,190 for 4,404,000 board feet of timber. Throughout the decade, Menasha sold dozens of stumpage sales to the local mills in Oregon and Washington as summarized below:

	1970–1980
Timber volume sold	438,000 MBF
Profits	\$36,000,000

By the 1970s, the majority of the main road systems on Menasha timberlands were well established. For the most part, logging was now done "uphill"—that is, the logs were dragged up the hill by large mobile yarders and then loaded onboard a log truck for their downhill ride to the mills on the Bay, or out the "back door" of the Menasha tree farm to mills in the Willamette Valley. This new logging system was very different from the days of downhill yarding with steam- or diesel-powered donkeys, which in turn were far more versatile and powerful than the old oxen teams.

The equipment that made up a given logging operation during this period would include, for example, a big Skagit BU-98 yarder, a Caterpillar D-8 bulldozer and a Link Belt 8400 shovel (loader). The cost of this equipment could well exceed \$1 million.

As Menasha was still harvesting large and valuable trees on its Oregon coastal tree farm during the 1970s, it began requiring the loggers to use a new timber-falling techniques developed by Weyerhaeuser on its Millicoma Tree Farm in Coos County. This process involved using hydraulic jacks (called Dellwood Timber Tippers after the company's logging operation in the Coos River drainage) to tip the timber up hill in order to minimize breakage. Safety was a major issue with the Timber Tipper, because the fallers were now below the falling tree and would have to scramble to insure that gravity did not send the base of the tree in their direction.

Because Menasha sold its timber in the form of cutting rights on specified parcels of land, the job of company foresters focused on ensuring four main things:

- That every usable log was removed from the logging site
- That contract loggers had adequate fire equipment at each logging site in case a fire developed on their operation

- That Menasha was promptly paid for the logs removed
- That the land was properly prepared for reforestation.

A total of ten employees made up the timber staff of Menasha during the decade of the 1970s.

Veneer

The success of its timber selling operation did not stop Menasha from exploring new ways to generate profit. One important effort would involve the making of wood veneer. It started in 1969 when Everett Doyle, President of Doyle Milling Co. Inc., sold a portion of his interest in a veneer plant in Myrtle Point, Oregon, to Menasha. Although its plywood plant had been dismantled just two years earlier, Menasha management judged that having a small converting operation connected with its timberlands would be advantageous. Within a few years of the initial partnership, they constructed an additional veneer plant near Coquille, Oregon.

The original mill was designed to peel logs eight feet long and eight to sixty inches in diameter with the resulting veneer products sold on the open market. Timber from the Hudson Ridge unit of Menasha's Oregon tree farm was set aside for supplying logs to this mill. As with the Menasha-Coos Head Plywood operation in 1950, log transfer pricing would again surface as a major source of conflict between the timberlands and the mill operations.

The other mill built by the partnership at Coquille, Oregon, was designed to handle very small-diameter logs. It, too, was a veneer plant with the output also sold into the open plywood market. The new mill also contained two small "box shook" lathes that manufactured small pieces of veneer eighteen to twenty inches in length, four to six inches wide, and a quarter-inch thick to be used to make wooden crates for the grape and lettuce business in California.

The partnership dissolved in 1974 but both mills continued to operate under other owners. Everett Doyle died in 2002. As a new panel product (oriented strand board) began eroding the plywood markets in the west, these mills slowly fell into disrepair and today they are no longer in operation.

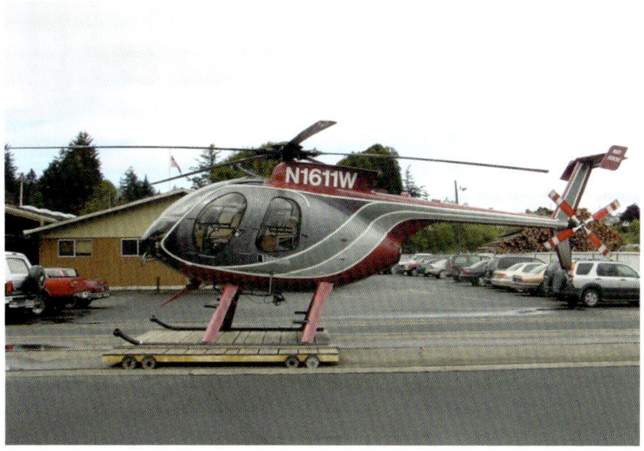

Menasha's Hughes 500-E, 1990–present

Helicopters

Since the buyers of Menasha stumpage paid for only those logs that were removed from the logging site, it was the job of Menasha foresters to carefully inspect each active logging area for missed logs. A single log might be worth $200 to $300. Given the steep topography and heavy brush cover on the logging area, several logs could be overlooked by the logger—and also missed by even the most careful Menasha inspector.

It was Everett Doyle who first suggested that Menasha consider using a helicopter to conduct the logging inspection.

The author recalls vividly the work involved in analyzing the economic advantage of owning a helicopter as part of Menasha's timber operation. The cost/benefit ratios were computed from every conceivable angle: How many man-hours would be saved? How many gallons of vehicle fuel, employee hours, and maintenance costs would be saved versus the fuel and time spent driving to the logging site and slogging through the mud and brush looking for the derelict logs? Would Menasha foresters spend more time preparing future logging operations and conducting forestry projects? The questions went on and on.

Menasha's Hughes 500-C (1984–1990) model is seen here dipping water from the Coquille River to assist with the fighting the Bob Cat Ridge wildfire on Menasha timberlands in 1987

Menasha's Hiller 12-5 helicopter went into service in 1975

The crash site of Menasha's Hiller 12-E helicopter in 1984

The decision kept getting postponed until one morning in 1974 when the author, then living in Neenah, Wisconsin, and working at the Menasha corporate headquarters, received a call from Senior Vice President Mowry Smith Jr. He said, "Bill, you go right in to Dick Johnson's office [president of Menasha at the time] and you tell him that we need to buy the helicopter and that he must approve the expenditure authorization for $125,000. Now."

The author was twenty-eight years old and, with considerable trepidation, followed Mowry's orders. Menasha purchased its first helicopter. The Hiller 12-E went into service in June 1975.

Over the years, helicopter use has broadened to include work both inside and outside Menasha. The external contracts have included some unusual projects: the harvesting of Christmas trees from the orchards around Corvallis, Oregon; lifting small refrigeration units to the top of buildings; surveying birds on the rocks along the coast line; darting and transplanting full-grown Roosevelt Elk; and, most important, fighting forest fires. The breadth of use that would evolve from having a helicopter well exceeded even the most ambitious economic justification back in 1974!

A near-fatal accident happened to the first Menasha helicopter in 1984 as it was on a routine reconnaissance mission with two company foresters on board. The ship lifted off from an abandoned logging landing to inspect a patch of timber near Schofield Creek north of Ten Mile Lake in Coos County. Following the normal lift-off procedures common to helicopters working in steep topography, the ship sped downslope to gain air speed and elevation. A nearly invisible power line cable hooked the toe of the skids and the ship pitched forward. The main rotor blade severed the tail boom, disengaging the rear tail-rotor gyro. The helicopter plunged through the forest canopy. The Menasha pilot, one of the best the author has ever known, had the presence of mind to turn off the electrical power (the magnetos) and did his best to fly the falling ship down through the trees.

While the helicopter was a total loss, the pilot and the two company foresters on board sustained serious, but not life-threatening injuries.

At his retirement party twenty years later, the pilot, Ron Patton, recalled with emotion cracking his voice that while he was lying in a hospital bed after the crash with a broken back, the author made the following comment to him: "Ron, as soon as you are able, I want you out hunting for a new helicopter for us to buy, and I want to be the first one to fly in it with you."

The company purchased two more helicopters during the next twenty years: a Hughes 500-C model, and the current ship, a Hughes 500-E. Each had its unique capabilities, but the work was always directed as originally planned—the ships were never used for joyriding.

In addition to its other duties, today's Menasha helicopter is used to traverse land boundaries from the air to obtain an accurate description of the unit for record-keeping. In order to map the area, an antenna equipped for Global Position System (GPS) is used. The helicopter flies the perimeter of the area with the GPS unit engaged. The information is downloaded into the company computer creating a map that accurately defines a unit boundary, road location, or any other two-dimensional area.

Over the years, the company helicopter has been used for a wide variety of outside contract work in southwestern Oregon. It is often called upon by the local drug enforcement agencies to fly their personnel in a surveillance of areas where marijuana is suspected of being grown. It might also be found with a wildlife expert onboard doing Roosevelt Elk surveys within nearby watersheds.

Timberland Acquisition

During the 1970s Menasha purchased an additional 9,223 acres of timberland to add to the tree farms for a total cost of $4,372,000, and sold 1,124 acres deemed surplus to the operation.

By 1980, the Menasha inventory of land and timber in Oregon and Washington was as follows:

	ACRES	TIMBER VOLUME
Oregon	82,677	713,247 MBF
Washington	12,294	45,647 MBF
Total	94,971	758,894 MBF

A typical mobile yarder logging setting on Menasha timberlands in the 1970s

Salmon Farming

An interesting sideline business developed in the mid-1970s, unrelated to the timber operation and only incidentally connected to Menasha's North Bend paperboard mill.

Several fish biologists approached Menasha with a proposal. They wanted to lease a small tract of land adjacent to the paper mill on the east side of Jordon Cove on Coos Bay and build on it a salmon release and recapture operation. Their idea was to make a profit from rearing, releasing, and recapturing Coho and Chinook salmon. Their company was called Anadromous, Inc. Ernie Lewis was its president.

After Menasha agreed to lease the requested land to Anadromous, construction of the "salmon ranch" began. The operation started in a rather meager way with holding ponds for the fingerling salmon scooped out of the sand and lined with heavy black plastic visquine. Once this process was completed, the small fish were dumped into the ponds.

The first problem was the feeding frenzy that evolved as soon as the seagulls got word that there were concentrated salmon fingerlings ready for the taking. But that was not the last difficulty.

One thing led to another and the scientist, having run out of money, approached Menasha in hopes it would invest in the venture, which we did by taking a 51 percent ownership in the salmon ranch. Since it took two to three years for the salmon to mature after being released, Menasha found itself in a "wait and see" mode.

A couple of years later, another salmon ranch was established in the lower section of Coos Bay in competition with Anadromous. Now the plot thickens. Salmon are attracted back to the smell of the area from which they were spawned (traditionally a stream, but in this case a visquine-lined sandpit). For some reason—perhaps because the tide ebbs and flows past each of these salmon ranch facilities every six hours—only the "brightest" fish seemed to find their way home. Then the two salmon ranches on the Bay battled over whose fish belonged to whom.

Menasha sold its interest after a couple of years of poor returns, and both salmon farms on Coos Bay were eventually abandoned.

A bomber drops a load of fire retardant slurry on the Camas Creek fire in 1976. Note the steel yarder that is the target of the bomber. Unfortunately, the fire consumed the machinery—but the good news is that no one was hurt

Douglas-fir seedling

Forestry

During the 1970s Menasha was at the forefront of forestry practices in the Pacific Northwest. The company was on course for growing the maximum amount of marketable fiber possible on each of its timberland acres. Forest inventories were regularly taken on the property to check on the progress of the crop of trees.

Reforestation of the recently harvested units remained the main priority of the forestry staff. No longer did the foresters rely on Mother Nature to produce a bumper crop of seed in the forests adjacent to recent logging operations, nor was aerial seeding the appropriate method of predictably creating the next forest crop. Hand planting of seedlings from selected seed stock was the mode of the day. Seed from the Menasha-owned genetics orchard was beginning to become available to company foresters, adding superior growth characteristics to the new forests on company lands. In general Menasha was planting five seedlings for every tree harvested, an annual tree planting effort of one million trees per year—and that's a lot of holes to dig in the winter. This program would take more than three months each year to complete. But good results were quickly apparent.

Following the tree planting project, the new forest was tended like a garden. Weeds were controlled to ensure that each seedling had the optimum amount of sunlight and nutrients. Insect and animal damage was inspected regularly and proper action undertaken to prevent further loss. After the trees reached a height of ten feet, they were judged "free to grow" and the foresters' attention was redirected to the next area of young seedlings.

But of greatest concern to all foresters is wildfire. While Menasha has been spared the ravages of a large forest fire on its lands, wildfires do occur.

Today airplanes and helicopters are used in conjunction with scattered lookout towers to alert fire agencies to smoke in the forest. The Menasha helicopter has been used extensively to hold small forest fires at bay until ground equipment and crews can get on-site. During the summer months, the ship remains on standby to be ready to depart at a moment's notice.

A fully loaded ocean freighter departing Coos Bay with Menasha logs bound for the Chugoku sawmill in Japan

chapter
ten

The International Period: 1981-2003

Following the very successful decade of the 1970s when the demand for Menasha timber in Oregon and Washington was strong, both demand and prices "hit the skids" in 1979–1980 and no one wanted to buy any Menasha timber. Why? In hindsight, three macro-factors came together at once:

1. The federal government tightened the nation's money supply in order to control runaway inflation. This action in turn drove interest rates through the roof (and new housing construction into the basement).
2. Housing demand for the 77 million baby boomers had peaked in 1979 and the demand for new houses evaporated. No one was building homes.
3. With falling prices for lumber, chips, and plywood, the log and timber market plummeted. This created a real dilemma for the agencies of the federal government who sold timber contracts to companies throughout the West. Now that essentially all forest product markets were down, many of these public agencies' (US Forest Service, Bureau of Land Management, and State of Oregon) timber sale contracts were expiring. Should the government forgive the contracts and resell the timber, or make the purchasers honor the contracts and go bankrupt? The government chose a middle ground. It deferred the contract termination dates and released many purchasers from some of their contract obligations. A new dynamic was about to unfold in the timber industry in the West.

Ken Ann Maru

Concurrent with these changes, several Menasha executives retired: Dick Johnson as president and CEO, Al Stinchfield as vice president of the Land and Timber Division, and Mowry Smith Jr. as senior vice president. New leadership was handed to Donald C. "Tad" Shepard (a grandson of C. R. Smith) as president and CEO of Menasha, and Bill Lansing (the author) as vice president of the company's forest products operations in the West. The company's board of directors judged they were ready for the new assignments.

The new management team, however, was greeted with an incredible natural disaster: the explosion of Mount St. Helens on May 18, 1980.

Donald C. "Tad" Shepard

Bill Lansing

The following account, written by Randle, Washington, residents Barb and John Ensminger (a Menasha forester) later ran, along with the photos they took, in the company's newsletter:

> *Mt. St. Helens lies only twenty-three miles southwest of our home and after the initial eruption on March 27, 1980, we both became fascinated by the mountain. We had gotten a light dusting of ash from the first few eruptions. Now, there were reports the mountain had started expanding at a rate of five feet a day. The scientists knew*

The explosion of Mount St. Helens, May 18, 1980

something was going to happen, but what or when no one knew. The mountain was on everyone's mind.

May 18, 1980, was a gorgeous day in Randle. The skies were clear and it looked like a great day to finish planting the garden. Barbara was washing dishes when the plume came into view. The boiling clouds were moving at an incredible speed toward our house. We grabbed the camera and ran outside to get some photos. The speed of the plume was overwhelming. The plume was overhead and traveling to the N. Within minutes strong winds started blowing from the direction of the mountain. The sight of what was heading towards us was awesome.

We decided to go to Morton and take more photos, not fully realizing what was going to happen. By the time we reached Highway 12, approximately 4.5 miles away, the sky was starting to darken and the wind had picked up even more speed. We stopped at a little store to buy film, and as I walked in, the lights flickered out. The sky beyond the store was pitch black, and the darkness was advancing on us rapidly. When the globs of ash started pelting the car, we decided it was not time for pictures. As we started back home, the wet ash was covering the windshield, and we had to use the windshield wipers. It was like mud. Then the air became filled with powdery ash. Lightning was striking and visibility was zero. The inside of the car was filled with ash and it was starting to smell like sulfur. It took us thirty-five minutes to drive what had taken us ten minutes before. Once home we covered our heads with blankets and ran back into the house. We tried the radio, but there was no news. We never did hear any information until 1:00 p.m. (five hours after the explosion). The media was terribly slow on this one. It remained dark until around 12:00 noon.

When the darkness finally ended, the sky was an odd yellow color. The air had a strong sulfur odor, and we could hear the mountain rumbling in the distance. The cows were like gray humps in the pasture and an occasional moo was heard from their direction. Two or three inches of ash were everywhere, and everything was gray. There was hardly any color anywhere; just the dull gray color of the ash and the odd yellow sky. All that day we could hear the mountain rumbling.

Things have slowly improved, but not completely. Our garden did grow, even though we finished planting it with masks on our faces. But our driveway is still a slimy mess of wet ash right now (January 1981). My clothes are full of ash after I have been in the field (inventorying the effect on Menasha's timberland). Our windows in the house are still taped closed. . . we even had to wash our Christmas tree. The ash adheres to everything, and it does not rinse off easily. We had to clean the Christmas ornaments when we put them away.

We are no longer apprehensive when we go somewhere, but we don't go near the mountain. The beautiful forest where we first went to collect samples of volcanic ash off the snow pack has been totally devastated. Our ears always perk up when new seismic activity is reported. We are still curious about the mountain, but with added respect for the power of nature. The full impact of the potential dangers of the volcano never really came home until a friend in Oregon asked us if we had some suitcases packed and ready in case of an emergency evacuation. One wonders, how bad will it be next time? Last time it completely devastated areas within eight miles from our home.

Volcano-caused timber destruction in the Clearwater drainage on the northeast side of Mt. St. Helens (Courtesy of Weyerhaeuser)

The news media peppered the airways with daily warnings of the mountain's activity but it was still hard to grasp the massive scale of the disaster. The author recalls talking to the spouse of Menasha's Washington forester after the initial eruption. The two of them lived outside the small town of Morton, Washington, some twenty-five air miles from the mountain, in the center of Menasha's tree farm. The forester's wife said she saw a huge gray-black cloud rising over a tall ridge that separated them from a direct view of Mt. St. Helens. Shortly after, pumice stones the size of marbles began pelting their yard.

Its destruction was monumental in many ways: Lives and property were lost, and no one knew what effect the blanket of volcanic ash would have on the thousands of square miles of timberland adjacent to the mountain.

Weyerhaeuser Company logging equipment destroyed in the Mt. St. Helens explosion (Courtesy of Weyerhaeuser)

While the devastation was incredible on the Weyerhaeuser company timberlands in the area, the Menasha timberlands in Washington escaped major damage, receiving only a covering of an inch or less of fine volcanic ash. However, no one in the forest industry had ever experienced this before and scientists could only speculate as to what the ash might do to the green forests.

As it turned out, the ash had little impact on the forests surrounding the mountain. The Weyerhaeuser Company, with a significant amount of their local forests literally flattened by the explosion, took a leadership role in harvesting the damaged timber and reforesting the landscape. The company is to be complimented on taking a natural disaster and converting it into a thriving new forest. John Wilkinson, a current director of the Menasha Forest Products Corporation and then a vice president for Weyerhaeuser, was in charge of the rehabilitation project.

Welcome to the 1980s

After ten years of producing average annual profits of $3.6 million, Menasha timber operations in 1981—for reasons mentioned at the beginning of this chapter—reported a loss of $196,000. This was not a good start for the new management team!

Menasha's Raul Diaz (new product development manager), Bill Lansing (vice president, general manager, and author of this book), and Frank Lyon (timber manager) stand on the Great Wall of China, steeling themselves against the bitter cold wind blowing in from Mongolia

As things turned out, this was the low point in a very successful period for the company. It spurred new and profitable efforts that would take Menasha around the world. The new era began in 1981, when three Menasha timber managers from the North Bend timber operation traveled to China, Korea, and Japan in an effort to find new markets for Menasha logs.

Many barriers were encountered in developing international markets for Menasha logs, ranging from language misunderstandings and cultural differences to business protocol challenges and currency exchange issues. In addition, the markets required different lumber sizes, wood went to different uses than previously experienced in the US, log measurements standards were different—and of course there was some odd-tasting food.

Frank Lyon, Menasha vice president and timber manager, discusses future log business with Mr. Y. Horikawa, owner of a large sawmill in the Chugoku Prefecture in southern Japan, in 1986. Shoichi (Stanley) Yoshino of the Nichimen trading company helped translate

The Menasha timber group persisted in developing the Asian market and soon achieved success. The first company cargo of export logs was shipped from Coos Bay on the vessel *JAVARA* on June 18, 1982, destined for the port of Qinhuangdao in the People's Republic of China. It carried $1.9 million worth of Menasha logs onboard!

The author recalls the ship being of "rust-bucket orange" in color, with antiquated cranes and a Filipino crew. Its last cargo was coal shipped from Chile. Apparently, it had a limited history with logs, but it was the largest log ship ever to come to Coos Bay at that time.

While every effort is made at the logging site to manufacture each log to the customer's specification, Menasha quickly developed its own log merchandising process for insuring maximum quality control. Here a Menasha log yard technician Loren Osborne improves the quality of a log destined for a sawmill in Japan. By the mid-1990s Menasha had become the fourth largest US log exporter to Japan

That was the first year of a very successful business venture for Menasha's western timber operations. From 1980 through 2003, Menasha would harvest 1.3 billion board feet of timber from its lands in the Pacific Northwest and generate more than $240 million in profits.

Menasha quickly adapted its systems to the log export business in East Asia. Logs had to be assembled in sufficient quantities and specific qualities to completely fill an ocean freighter. This usually took three to four months to assemble the logs at the water's edge, and one week to load them aboard ship. In the early 1980s, a full cargo of Douglas-fir logs destined for China would have a sales value of $1.5–$2 million; the customer was 6,500 miles away— but never once did the company fail to be paid for its cargo.

CB Cedar

In 1987, the Menasha management in Coos Bay recommended that a small local sawmilling operation would be an attractive asset complimenting the timberlands in the region. The effort was directed toward red cedar, a lumber product used for decking, posts, and decorative fences. Since the Menasha management knew little about sawmilling, a partnership evolved with the Caffall Bros. Forest Products Company in Wilsonville, Oregon.

The partnership was known as CB Cedar. The mill cost $4 million to construct, and many cost overruns were encountered. The mill started in 1987 and quickly ran into major problems:

The CB Cedar sawmill was located in the building positioned parallel to Isthmus Slough. The large building perpendicular to the sawmill contained the green chain and planer facilities. The buildings and property were leased from Georgia-Pacific and was the site of their plywood mill in the 1970s; before that it was the site of C. A. Smith's "Big Mill" and the Coos Bay Lumber Company

Only a minor portion of the logs needed to supply the mill were to come from Menasha timberlands. The majority of the red cedar logs were planned to originate from the purchase of logs being harvested by other timberland owners in the region—including Bureau of Land Management and US Forest Service timber sales.

The policies of the federal government in Washington, D.C., regarding timber-selling programs went into a deep spiral resulting in an ever-decreasing amount of timber offered for sale in the local area. Menasha was trapped in a supply squeeze as this expected source of red cedar logs vanished.

The quality of red cedar lumber cut from the local logs in southwestern Oregon contained excess amounts of water, making freight costs non-competitive.

Local red cedar trees were also susceptible to attack by an insect that left small cavities (called larval galleries) in the wood. The cavities were not a structural problem but posed a major cosmetic problem that could not be overcome.

After a year of operating the mill, the partnership was dissolved and the facility sold to Weyerhaeuser in 1989 for $3.5 million. Weyerhaeuser converted the sawmill to cut metric-size lumber for the Japanese market. The mill ran successfully for several years, but fell upon hard times as the Japanese economy began to falter in the mid-1990s.

Forestry Consulting

By the mid-1980s the results of Menasha's intensive forestry programs were evident, even to the general public. As an acknowledgement of Menasha's dedication to timberland stewardship, many neighbors throughout Oregon and Washington asked company foresters how they might get such superior-looking young forests on their property. This led Menasha into a forestry consulting business in order to assist the small non-industrial forest owners in southwestern Oregon to attain the same good results on their timberland properties.

The professional consulting advice was often followed by a request from the neighbor for the Menasha field crew to perform the on-the-ground implementation of the plan. These requests led to a new Menasha business started in 1984. The subsidiary was called Forest Management Services (FMS).

Log trucks are driven onto the dock alongside the vessel and logs are lifted directly from the truck and loaded aboard the ocean freighters in 25,000–30,000 pound units. A full cargo of would contain 5–6 million board feet or approximately 23,000 individual logs

Higher and Better Use

Not all lands owned by Menasha are suitable for long-term investment as a tree farm. Some of the properties are simply not strategic to the long-range tree farm goals, while others have significantly higher value for other uses. For instance, land more suitable for rural homesite development is separated from the tree farm and disposed of as surplus property.

Another subsidiary, the Menasha Development Corporation (MDC) was formed in mid-1984 to handle the disposal of these surplus lands. MDC sells these surplus company lands in the most efficient means available. This might entail a wholesale transaction where a large block of land is sold, or a parcel of surplus land might be subdivided into smaller lots for a retail sale.

Timberland Acquisition

From 1981 to 2001, Menasha acquired several small to mid-sized timber tracts located within the same operating territory as its Oregon and Washington holdings. During this timeframe, a total of 19,901 acres of timberland containing 150 million board feet of timber were purchased for $40 million, 10,342 acres were sold, a total of 1.9 billion board feet of timber was harvested from the company's tree farms, and profits of $213 million were generated. While the company did not have a clearly stated goal to increase the size of its timberland holdings out West, considerable cash flow was being generated to the point where the author was able to execute the deals without much fanfare—or permission!

The log export ship Aeries Harmony *loading logs at the Menasha log-sorting yard in North Bend, April 2004, for a sawmill customer (Chugoku Mokazai) in Kuri, Japan*

The bark of this tree was stripped away by black bears, a species whose range includes some of Menasha's prime forests in Oregon and Washington

The porcupine in this photo has stripped the bark from a young tree—a sight seen at several locations at Menasha tree farms. These trees will not survive

BEARS, BEAVERS, AND PORCUPINES

Many biological factors impact a tree throughout its life—some good, some bad. One such unsuspecting, but well known, animal can have deadly consequences to a Douglas-fir tree: The black bear

When a black bear comes out of hibernation in the spring, it has a sugar imbalance in its system. Licking the inner "sap" of the Douglas-fir tree can cure that chemical deficit. This is a learned behavior passed along from mother to cub(s). Trees stripped in this manner will eventually perish. Menasha foresters have found a non-lethal way of dealing with this localized bear problem—they feed them!

Menasha foresters also control the bear population by allowing hunting on company lands. The fish and wildlife professionals counsel Menasha foresters as they determine the carrying capacity of a specific watershed. The goal is to provide a healthy habitat for the animals, yet protect the young forests as they mature.

Here a Menasha bear-feeding station contains pellets laced with the same chemicals that the bear seeks from the sap of a tree

When a newly harvested area is replanted with genetically improved seedlings, the foresters face another problem: The young seedlings are a preferred food source for an animal nicknamed the mountain beaver. These unusual rodents—rarely seen although the damage they can cause is very evident—are not a true beaver, but smaller burrowing animals that resemble large gophers. They come out only at night to feed. They are also highly prolific and unless the population on a newly planted site is reduced, the entire planting effort will be for naught.

Once the seedlings reach five to six feet in height, the trees are too large for the mountain beaver to do any real damage.

Although you might think that once the planted seedlings reach beyond the size where the small rodents can destroy them or competing brush can choke them out, the foresters job of protecting the trees from further damage is done. Not so. The next culprit that will gnaw away at the tree is the intrepid porcupine.

Porcupines attack trees that are about ten years of age. When young, the trees' bark is relatively thin and an abundance of sugar is produced in their sap. Porcupine damage can usually be seen on the stem of the tree some ten to fifteen feet above the ground. The incisor teeth and sharp claws are perfectly designed to strip away the bark and expose the food source beneath. The impact on the tree is the same as that of the bear: If the porcupine strips the bark completely around the stem, the upper part of the tree will die and produce a multi-topped mature tree of little value for lumber.

Not all animals in Menasha forests damage the trees. However, the mountain lion population within Oregon's forests has grown and shifted to areas where their food source is more plentiful: The new forest plantations where elk and deer are abundant. As the timber harvesting philosophy of the federal government has shifted toward a no-cut program, the species dependent upon clearings concentrate on private forestland where intensive forest management is used. This in turn brings the carnivores into closer proximity to human populations—and clashes do occur.

A bobcat in a tree on Menasha's Oregon forest lands

Mountain lion in a tree on Menasha's Oregon forest lands

By 2000, the Menasha inventory of land and timber in Oregon and Washington approached 100,000 acres as follows:

	ACRES	TIMBER VOLUME
Oregon	82,639	765,978 MBF
Washington	13,945	88,166 MBF

By the mid-1980s both export and domestic (US) log prices were exceptionally strong and predicted to get even stronger. In 1984, Menasha launched its program of purchasing timber harvesting contracts from other timberland owners in the Pacific Northwest (non-fee timber contracts) as well as purchasing truckloads of logs delivered by loggers harvesting timber from other sources.

The new additions to the Menasha tree farms discussed above, combined with the large quantity of non-fee timber sale contracts available, helped the company quickly grow to become the nation's fourth largest log exporter to Japan. As Menasha matched its manpower to the growing export market with these non-fee contracts, the overall organization grew to more than 100 people as offices were opened in Idaho, California, and Alaska.

This strategy proved very successful until 1997, when Asian banking crisis began to take its toll on the Japanese economy, resulting in a plummeting demand for logs and other forest products.

It took a couple of years for the local management of Menasha to recognize that this change in the global market for timber was not an anomaly, but something that would have long term consequences for those who sought new Asian customers for their logs. Menasha began to shrink its manpower starting in 1998 and when the final count was in, 50 percent of its valued employees were terminated.

Large rocks stretching across salmon migration streams on Menasha tree farms help catch small gravel to improve spawning

Gravel bars improve the success rate of salmon spawning on stream coursing through Menasha ownership in Oregon and Washington

Heavy equipment is needed to place large boulders in the streams to improve salmon spawning opportunities

The watercourses throughout Menasha tree farms are precious commodities and special care is exercised to insure that the water remains clear

Spawned-out Chinook salmon in a stream running through the Menasha tree farm in Oregon

The cross-section of six-inch Douglas fir tree on the left (forty years of age) is an example of tree growth in an unmanaged young forest, while the block on the right (seven years of age) is from a Menasha managed forest

In the spring, plastic bags are used in the Menasha seed orchard to seal the female cones on the Douglas-fir trees from being fertilized by naturally occurring pollen floating in the air at the time of fertilization. In this way, pollen from another genetically superior parent can be injected inside the bag to insure the proper genetic crosses are made. The trees in the orchard are pruned to allow for easy access to the cone-producing tops of each tree

TODAY'S FORESTRY

Intensive forest management has now become second nature to Menasha. Effective silvicultural practices (forestry programs designed to increase and improve the amount of wood fiber grown) are employed on each acre to produce the greatest long-term economic return while at the same time protecting the fish and wildlife that co-exist in the working forest.

The process begins at the Menasha genetic seed orchard where the very best genetic material localized to the company tree farms is managed. Crosses between superior performing parents are made with the expectation of their offspring increasing forest survival and growth by 20 to 25 percent above natural levels.

Once the cones from the seed orchard are ripe, they are harvested, dried, and the seed extracted. As with the less intensive forestry practices of the 1950s and 1960s, the seed from the orchard can be stored in freezers for several years before their viability deteriorates. The cost of this "super seed" is about 100 times the cost of the seed used thirty years ago, but the results are worth the expense.

As company foresters plan upcoming logging operations, they also determine the amount of seedlings they will need to reforest the areas immediately after harvest. The exact amount of seed is then removed from the freezer storage and delivered to a contract nursery where the germination process begins.

As with all agricultural crops, the land needs to be prepared for the next crop. Since there is considerable brush and logging debris (limbs, tree tops and other material) left after the harvesting operation, the potential for wildfire increases. Foresters accomplish a dual objective when the fuels on recently harvested sites are eliminated—they reduce the risk of a wildfire and the land is better prepared for the tree planting effort. Fire, in a very controlled situation, is used to accomplish these objectives.

After the land is prepared, the final stage in harvesting the forest is to plant a new crop. The following series of photos shows the reforestation of a Menasha clearcut site near Greenacres, Oregon, over a twenty-five-year time span.

The goal of intensive forestry (silviculture) is to grow a consistently superior tree on each acre of Menasha lands that will produce uniform products for the homebuilder. Like the farmer, a forester's goal is to grow the optimum amount of wood on each acre of land under his/her care. Foresters have learned that the application of these scientific programs will produce the desired financial results for the landowner.

While Menasha is clearly focused on the financial returns from its investments in trees, it realized early on that it held an equally important responsibility to the stewardship of the land, animals, and water. Riparian zones, for example, are critical to the ecology of the of the stream environment as well as the forest lands further up the slopes. Menasha takes great pride in the work it does to enhance these environments by proactively managing them.

As a responsible forestland owner, Menasha, through its forestry practices, also give consideration to the other resources on its tree farms.

Douglas-fir cones and seed from Menasha's Oregon seed orchard

After the cones are picked from Menasha's genetic seed orchard and the seed extracted, the seeds are germinated in a climate-controlled greenhouse where they may stay for three to six months, until they reach a specific size. This photo shows the IFA Nurseries, Inc., greenhouse forest nursery in Nisqually, Washington. Menasha is a stockholder in the company. Here, IFA President Dave Stearns displays some Menasha seedlings

Clearcut operation on Menasha lands, 1979

Same clearcut five to seven years after logging

The Menasha helicopter igniting logging slash in a manner that controls the fire and produces a smoke column that will not intrude into populated areas. The process not only reduces the chance of a natural wildfire, but prepares the planting site, mimicing nature's technique

Same clearcut fifteen years after logging

Same clearcut twenty-five years after logging

One of Menasha's sustainable tree plantations

From the controlled climate of the greenhouse, the seedlings are transplanted into a nursery bed for another six to twelve months before they are ready for outplanting on a newly prepared Menasha logging site. This photo shows the IFA Nurseries, Inc,. field nursery at Toledo, Washington

What is left of the old Menasha Plywood Corporation building lies to the right of the dock behind Menasha log decks. The land is now a log sorting and export facility operated for MFPC by Jim Lyons, grandson of James A. Lyons of the Irwin and Lyons/ Coos River Boom Company

chapter eleven

Into the New Century

On April 1, 2001, Menasha spun off its west coast timber business into a new stand-alone company called Menasha Forest Products Corporation (MFPC). This new company remained in private hands, owned by the family descendants of Elisha D. Smith and Julia Ann Mowry who also retained ownership of the Menasha Corporation. The spin-off transaction was complex, in great part because complete legal title work had to be determined for every parcel of land that would eventually end up in the ownership of the new company. It took more than two years to complete the spin-off requirements, ending in the receipt of a Private Letter Ruling from the Internal Revenue Service stating that the transfer of assets to the new company could occur tax free. Both the executives of Menasha Forest Products Corporation and its parent company (assisted by the counsel of Arthur Anderson and three law firms) worked together to complete this transaction. The headquarters of the new company remained in North Bend, Oregon, and occupies the same building built for the Menasha plywood operation in 1963.

To the outside world of customers, suppliers, and the communities where its employees and families live and work, MFPC remains Menasha, the company that started business in Empire City in 1905, purchased the assets of the Southern Oregon Company, and built the plywood and paper mill in North Bend. Menasha's long history in the area is even reflected in the location of the MFPC corporate office, near the old dismantled Asa Meade Simpson sawmill and shipyard and just north of the old Menasha Plywood Corporation plant.

One might wonder as to the reasons behind the restructuring of Menasha Corporation by spinning off the timberlands into a completely new and non-connected organization—the only common ground between the two companies being the almost identical shareholder base. Two major reasons can be clearly identified:

The focus of the parent Menasha Corporation had gradually moved away from the timber business into an eclectic group of generally unrelated business enterprises, where the customer base, supplier contracts, and operational processes were incompatible with a timber operation. Essentially the timber business of Menasha no longer fit the strategic direction of the larger company.

The executives of the timber business of Menasha Corporation recognized in the early 1990s that the future of the business would be destined to produce cash for the growth of these other corporate ventures into the future. Between 1981 and 2003, the timber operations provided more than 25 percent of the net cash flow of the entire company, delivering more than $250 million in pre-tax profits during the same period. This did not sit well with the timber staff, as minimal cash was being put back into the timber operations for growth, and lobbying efforts began in earnest by 1995 to separate the two companies.

The MFPC office, North Bend, Oregon, April 2004

Timber Jack 2628 feller-bunchers are used where the topography is not too steep and where trees are smaller than eighteen inches at the butt

The control panel of the Timber Jack 2628

To increase production and reduce safety hazards at the landing, mechanical de-limbers with computer-driven mechanical cutoff saws were developed. Here a Link-Belt 3400 stroke de-limber and a Caterpillar 225 hydraulic log loader (shovel) work on an MFPC logging site in Coos County, Oregon, 2004

Inside the cab of a Caterpillar hydraulic loader (shovel) operating in the Menasha log sorting yard, 2004

A third rationale behind the spin-off might be worth mentioning. A few years before the creation of MFPC, several major publicly traded forest products companies had split off their timber from their integrated manufacturing company. International Paper Company spun their timber into a separately traded company called IP Timber, and Georgia-Pacific made a similar move. Essentially Wall Street and the public shareholders of these integrated forest products companies were demanding a better return from these manufacturing operations, knowing the land and timber portion of each company's business was masking the poor performance of the primary (sawmills, plywood mills, and pulp and paper mills) conversion operations.

Logging Operations

Much has changed in Pacific Northwest logging operations during the past century-and-a-half. Steam replaced oxen power in the early part of the twentieth century, then internal combustion engines were quickly adapted to replace the steam donkey. The old "misery whip" and "bucking saw" were eventually replaced by the gasoline-powered chainsaw. The art of rigging a spar tree is an art only—something old-timers discuss and no one but competitors in lumberjack carnivals practice.

A Sikorsky S-64 sky-crane logging timber from a sensitive area

Today professional foresters—engineers with computers in hand—play an integral part in designing the harvesting and road construction processes to minimize the impact on the environment, yet continue to produce needed commodities for the public and financial dividends for MFPC shareholders.

As we move into the twenty-first century, logging has made a slow but necessary jump to the computer age. Logging equipment today requires highly trained operators who are not only familiar with the logging process and general log markets, but who are also able to read and understand the computer dashboards and joysticks required to operate their monstrous machines—without stepping foot outside of the cab.

Where the topography is too steep to use these ground-based log-handling machines, Menasha contract loggers still use portable "spar trees." These new yarders are much smaller than the bohemoths used to pull in the big old-growth logs, and can swing on their bases to position each turn of logs in a safe and orderly position.

Under special circumstances, MFPC might use a large-capacity helicopter to harvest timber in environmentally sensitive or difficult-to-reach logging sites.

Forestry for the New Century

While Menasha can be proud of its long-term history of its forestry practices dating back fifty to sixty years, newer methods applied today to each acre of MFPC timberlands are even more intensive and effective. Time, scientific advances, and political pressure have combined to change, improve, and modernize forestry practices. Still, the same general pattern is followed today on each harvested site.

Company computers track the growth progress on every parcel of company land. From the data maintained, Menasha foresters assess the harvesting opportunity of each tract in their annual office planning cycles.

After the young forest approaches maturity, seven to ten years before final harvest, it is fertilized to increase the production of wood fiber

Months of preplanning work and hiking through potential logging sites is the first order of business of the harvesting cycle. Other professional forest engineers use their knowledge of the tree farm and computer data to develop the road systems and lay out the survey boundaries needed to accomplish each operation. Other foresters carefully monitor the progress of the growth of the young plantations to insure that proper stewardship and opportunities are available for the next crop of trees.

The land is harvested by Menasha contract loggers under the careful eye of experienced company foresters. Today, the company owns no logging equipment at all.

As noted in Chapter Ten, the land is properly prepared after logging for the upcoming reforestation effort.

Genetically selected trees are hand-planted on each acre—four to five planted for each one harvested. Each tree is approximately twenty-four inches in height and two years of age at time of planting.

Each acre is carefully watched to insure that each tree is given the optimum potential for survival and success.

The cycle is repeated every forty to fifty years—again and again.

As one watches the debates over consumption of forest products in our country, the fact that trees are a renewable resource is often buried in the appendix of the literature that tries to convince the electorate that forest management is somehow a bad thing.

To the contrary, we know from scientific research that wood is by far the most environmentally responsible building product known to man—bar none. Steel, aluminum, plastics, and so forth are all consummate consumers of non-renewable resources. The processes involved in bringing them to the public produce a large amount of carbon gases—the very food used by trees in combination with sunlight to produce a product we call wood.

From the time the author joined Menasha in the summer of 1969, a total of $13.5 million has been invested in reforesting Menasha lands after harvesting. To insure each seedling has the optimum condition with which to grow, an additional $24 million has been invested in intensive forestry practices to improve the quality of the trees growing on each acre in the Menasha forest.

Menasha prides itself on using its forest lands for multiple purposes, not only for the benefit of its shareholders, but also for the public at large. For example, in 2004 Coos County successfully bonded itself to bring natural gas to the communities within its boundaries. A pipeline twelve inches in diameter and sixty-five miles long had to be installed to reach from Roseburg, Oregon, to Coos Bay. The major route that it generally followed was none other than the Coos Bay Wagon Road. On occasion, the route deviated across current Menasha timberlands, for which a right-of-way was granted.

A Madill mobile yarder, a hydraulic shovel, and a stroke de-limber work on an MFPC logging site

Tree planter

People First

In the three years since the spin-off from Menasha Corporation, MFPC has purchased 36,411 acres containing 270 million board feet of timber for a total cost of $65 million. During the same three-year time span, MFPC harvested 222 million board feet of timber from its tree farms and booked profits of $31 million.

Where Menasha's original ventures into the timber business in Idaho, Oregon, and Washington were in the hands of two or three people, the current MFPC operation comprises forty-eight professional foresters, engineers, surveyors, financial accountants, and a real estate specialist. The company motto for twenty-five years has been "People First, Product Second." It remains the cornerstone of the company's culture. This philosophy permeates the organization and continues to serve it well.

Much has changed in Coos Bay during the past 100 years. Today there is only one sawmill on the Bay; the area's sole log export sort-yard facility is operated by Menasha. Coincidently, that export facility is in the same location as the old Menasha-Coos Head Plywood mill built in 1949.

Recent Timber Purchases

One hundred years after C. R. Smith started the company's move west with the purchase of the Blackwell Lumber Company lands in Idaho, MFPC purchased the second largest single timberland parcel in its history—the lands once owned by US Plywood Corporation in Gold Beach, Oregon:

Timberland purchases by MFPC in 2003	
Gold Beach lands	28,000 acres for $53 million
Hancock lands	7,000 acres for $6 million
Other lands	3,000 acres for $7 million

The company's 138,000 acres of timberland ownership in Oregon and Washington are the backbone of the business today. Timber is harvested from the land with all the care and responsibility that the current law and ethical requirements dictate. No mills are connected to these lands. Menasha understands the obligation it has to the land and to its shareholders, who have entrusted these assets to a group of professionals.

Coos Bay is a community that has seen the good and the bad times over the past 100 years. The boom/bust of the coal mining operations in the early part of the century was followed by the steady growth of the forest products industry up through the mid-1970s. Since that time Coos Bay has lacked a solid industrial base and is searching for its next opportunities.

Throughout it all, Menasha has sustained itself and is a cornerstone in the economy of the communities that surround the Bay. Once again, as the company motto says, we put "People first"! Along with its professional stewardship of the forest lands, Menasha's economic success has allowed it to give generous financial support to sustain its local cities, schools, and citizens while donating countless hours of its human resources, even on company time, to local organizations.

Some examples:

The Dollars for Doers program where company employees give twenty hours of their time during the work week in the local schools in an effort to bridge the gap between the public schools and the employers in the area. After the completion of the required time in the classroom, Menasha gives $250 to the specific classroom.

Community Service Award for Teachers program establishes a mechanism for teachers to receive a "no strings attached" endowment of $2,000 for teachers who are actively involved in community development programs outside the realm of public education.

Getting Down to Business seminar where three dozen teachers are brought together by the company to meet with business leaders to listen to the needs of the employment community in hopes that a stronger bond is build between what teachers are teaching and what employers need.

Each year Menasha gives more than $75,000 to local community organizations that provide for the development of youth and the disadvantaged.

The Menasha Forest Products Corporation is heir to a legacy of timber ownership that is matched by very few families or companies in America. It is truly a unique story of "Seeing the Forest for the Trees," but one that will always satisfy the motto of "People First, Product Second."

Will the future of Menasha Forest Products Corporation be sustained by these principles?

With continued careful management, a strong commitment to the community, and a belief in timber's enduring value, I believe the answer is a resounding, "Yes!"

A twelve-inch natural gas pipeline is laid across a portion of Menasha timberlands in Oregon. The line is sixty miles in length and brought natural gas to Coos Bay in 2005

COOS BAY THEN AND NOW

The following series of color aerial photographs taken in April, 2004 show views of the Coos Bay area similar to those black and white photos taken a half-century earlier.

Cape Argo Lumber inventory and the river terminal fuel storage and dock at Empire, 1950

The site of the original Henry Luse sawmill in 1856 and Oregon Southern Improvement Company's "Big Mill" built in 1883 is at the center of the photo just beyond the white buildings and dock facility. None of the original buildings remain and the dock is unusable

Georgia-Pacific sawmill, 1950. The land was sold to Glenbrook Nickel for the import of nickel ore (University of Oregon)

A modern Georgia-Pacific sawmill is located at the site of the old C. A. Smith mill site built in 1908. Fred Meyer and Safeway stores now occupy the site of the old Evans Products mill. The property is now for sale

Weyerhaeuser sawmill and plywood complex, 1950 (University of Oregon)

The former site of the huge Weyerhaeuser sawmill and plywood complex is now owned by the Coquille Indian Tribe, which operates a casino and hotel facility on the property. In 2004, the tribe purchased the vacant Weyerhaeuser land north of the casino

Mill B 1950 (University of Oregon)

The land once occupied by Mill B, now empty, was purchased by the Coquille Indian Tribe from Weyerhaeuser in 2004

Central Dock, 1950 (University of Oregon)

The large white building in the center of the photo on the edge of the Bay is what is left of Central Dock. The dock fell into disrepair in the 1970s and the owners leased the facility to Kennecott Copper Company, which imported copper ore from New Caledonia for the US market. This venture failed and the property was repossessed by the county for nonpayment of taxes. Its future is uncertain

Coos Bay Pulp Mill, 1950 (University of Oregon)

Coos Bay Pulp Mill, 2005

source materials

BOOKS AND ARTICLES

Allen, Alice Benson. *Simon Benson: Northwest Lumber King.* Portland: Binfords and Mort, 1971.

Anderson, Steven. "The Biltmore Forest School Visits Oregon Operations." *Forest History* (Fall 2001).

Andrews, Ralph W. *Glory Days of Logging.* Seattle: Superior Publishing, 1956.

Baldwin, Ewart M. and John Eliot Allen. *Geology and Coal Resources of the Coos Bay Quadrangle, Oregon.* Portland: State of Oregon-Coos County Coal Survey, 1944.

Beckham, Dow. *Bandon By-The-Sea: Hope and Perseverance in an Oregon Coastal Town.* Coos Bay: Arago Books, 1998.

Beckham, Dow. *Stars in the Dark: Coal Mines of Southwestern Oregon.* Coos Bay: Arago Books, 1995.

Beckham, Dow. *Swift Flows the River.* Coos Bay: Arago Books, 1990.

Beckham, Stephen Dow. *Coos Bay; the Pioneer Period 1851–1890.* Coos Bay: Arago Books, 1973.

Beckham, Stephen Dow. *The Land of The Umpqua: A History of Douglas County, Oregon.* Roseburg, OR: Douglas County Commissioners, 1986.

Beckham, Stephen Dow. *The Simpsons of Shore Acres.* Coos Bay: Arago Books, 1971.

Blodgett, Richard. *Menasha Corporation: An Odyssey of Five Generations.* Lyme, CT: Greenwich Publishing Group, 1999.

Bureau of Land Management, *The Coos Bay Wagon Road, Recognition of an Historic Travel Route.* Coos Bay District of the BLM. May, 1996.

Case, George Baxter. *The History of the Port of Coos Bay, 1852–1952* (typescript thesis), Edinburgh, TX: Pan American University, 1983.

Cour, Robert M. *The Plywood Age: A History of the Fir Plywood Industry's First Fifty Years.* Portland: Binfords and Mort, 1955.

Cox, Thomas R. *Mills and Markets: A History of the Pacific Coast Lumber Industry to 1900.* Seattle: University of Washington Press, 1974.

Dodge, Orvil. *Pioneer History of Coos and Curry Counties*, 2nd. ed. Bandon, OR: Western World Publishers and Printers, 1969.

Douthit, Nathan. *A Guide to Oregon South Coast History: Traveling the Jedediah Smith Trail.* Corvallis: Oregon State University Press, 1986.

Douthit, Nathan. *The Coos Bay Region, 1890–1944: Life on a Coastal Frontier.* Coos Bay: River West Books, 1981.

Mahaffy, Charlotte L. *Coos River Echoes: A Story of the Coos River Valley.* Portland: Interstate Press, 1965.

Peterson, Emil R. and Alfred Powers. *A Century of Coos and Curry: History of Southwest Oregon.* Coquille, OR: Coos-Curry Pioneer and Historical Association, 1977.

Pierre, Joseph H. *When Timber Stood Tall.* Seattle: Superior Publishing Co., 1979.

Puter, Stephen A. Douglas. *Looters of the Public Domain: Embracing a Complete Exposure of the Fraudulent System of Acquiring Titles to the Public Lands of the United States.* Portland: Portland Printing House, 1908.

Robbins, William G. *Hard Times in Paradise Coos Bay, 1850–1986.* Seattle: University of Washington Press, 1988.

Robbins, William G. *Landscapes of Promise: The Oregon Story, 1800–1940.* Seattle: University of Washington Press, 1977.

Smith, Mowry, and Giles Clark. *One Third Crew, One Third Boat, One Third Luck: The Menasha Corporation (Menasha Wooden Ware Company) Story 1849–1974.* Neenah, WI: Menasha Corporation, 1974.

Smyth, Arthur V. "Millicoma: Biography of a Pacific Northwestern Forest." *Forest History Today* (Spring/Fall 2001).

Stein, Harry. *Pope Resources: Rooted in the Past, Growing for the Future.* Seattle: Documentary Book Publishers, 2003.

Wagner, Judith and Richard. *The Uncommon Life of Louis Jerome Simpson.* North Bend, OR: Bygones, 2003.

Walling, Albert G. *History of Southern Oregon, Comprising Jackson, Josephine, Douglas, Curry and Coos Countries, Comp. from the Most Authentic Sources.* Portland: A. G. Walling, 1884.

Williams, Richard L. *The Loggers.* New York: Time-Life Books, 1976.

The Settlement and Early Settlers of Coos Bay, Personal Sketches, Eccentric Characters and Historical Reminiscences. Written by a Pioneer Resident of the Bay. Marshfield, OR: Printed at the Coast Mail Book and Job Printing Office, 1879.

Additional articles of interest can be found in *The Timberman* magazine, 1911; various Coos Bay newspapers (The *Harbor, Times*, and *World*) of the period; and the City of North Bend's 1944-45 contest, *Early-Say Stories*, published by the Coos-Curry Pioneer Historical Association; and the *Menasha Log* internal company publication.

GOVERNMENT AND CORPORATE DOCUMENTS

Articles of Merger of Coos River Boom Company and Menasha Plywood Corporation
into Menasha Wooden Ware Corporation
Coos County Deed Records, Coquille, Oregon
Book 102, pages 216–219
Book 11 pages 395–400
Book 15 pages 213–225
Book 20 pages 87–90 and 529–552
Jones Oregon Stevedoring, Co. log shipping records, 1982
Menasha Corporate Records:
Annual Reports of President, Vice-president and General Manager July 23, 1919
Annual Report to shareholders July 23, 1919
Annual Report to shareholders 1920-21
Annual Report to shareholders June 26, 1922
Annual Report to shareholders August 10, 1926
Certificate of authority to engage in Business in Oregon, June 6, 1906
Internal correspondence:
June 30, 1936 (Carlton R. Smith)
September 5, 1917 (Charles McGuire)
Land and Timber Financial Statements 1947–1950
Minutes of Menasha Board of Directors:
April 20, 1916
February 11, 1916
June 3, 1904
September 26, 1916
Special Report of Thomas M. Kearney, President, February 17, 1919
Southern Oregon Company Annual Report 1918 and 1919, Arthur Young and Company
U.S. Coastal Survey Labors 1861-1865, pages 15-43

PERSONAL CORRESPONDENCE AND INTERVIEWS

Norm J. Wiener, correspondence 4/9/2004 (Miller Nash LLP attorneys at Law, Portland, Oregon)
Personal interview David Miller Sr. and Diane Miller 4/7/2004
Personal interview William D. Hagenstein 11/18/2004
Personal interview John Johnson 2/23/2004
Personal interview with Norm Miller 4/7/2004
Personal interview Mike Groben 2/23/2004
Personal interview Carl Newport 4/12/2004
Personal interview Mowry Smith Jr. 2/14/2004
Personal interview Ted Ellingsen 3/31/2004

HISTORICAL SOCIETIES AND COLLECTIONS

Coos Art Museum, Coos Bay, Oregon
Coos County Historical Society, Coos County, Oregon (Ann Donnelly, Director)
Cowlitz County Historical Society, Longview, Washington (David Freece, Director)
Douglas County Historical Museum, Roseburg, Oregon (Jean Mitchell, Curator)
Elisha D. Smith Library, Menasha Wisconsin (Jack Fry, Director)
Lewis County Historical Museum, Chehalis, Washington (Karen Johnson, Curator)
Pacific County Historical Society, South Bend, Washington (Bruce Weilepp, Director)
University of Oregon Special Collections, Eugene, Oregon (James Fox, Head)

WEBSITES

http://arcweb.sos.state.or.us/state/forestry/pics
http://bluebook.state.or.us/culture/history
http://www.ccrh.org/comm/cottage/primary/claim
http://collections.ic.gc.ca/heirloom_series/volume4/308-311
http://www.cooshistory.org
http://www.eraoftheclipperships.com/tallshipsofthecoosbay
http://www.ecobooks.com/books/railroad
http://www.ecobooks.com/railroad
http://www.drizzle.com/~jtenlen/walewis/logging-truck
http://gesswhoto.com
http://www.gen.bham.wednet.edu/irondon1
http://www.germanheritage.com.biographies/ntoz/weyerhaeuser
http://www.historylink.org/output.cfm?file_ID=1639
http://homepages.rootsweb.com/~1compton/sumner/SFA/places
http://www.landgrant.org/history
http://www.mnhs.org/places/sites/jjhh/aboutjames.html
http://www.mortondepot.com/sparkman
http://www.nap.edu/books/030905285
http://www.nplnews.com/toolbox/history/publiclandhistory
http://www.or.blm.gov/ForestPlan/O_andC/General
http://www.prcn/org/forestrymuseum/steam
http://www.publiclands.org/museum/story/story10
http://www.railserve.com/JJHill
http://www.steaminthewoods.com
http://www.sweet-home.or.us/history/logging-history/old-photos
http://www.theworldlink.com/photopages/shipwreck
http://www.geocities.com/riffefamilytree
http://voteview.uh.edu

index

John W. Catching
C.B.W.R.Co. Ex. "H"
C.B.W.R.Co. Ex. "H"
Jacob Hisley Oct. 17, 73
P.R.E. Wm. C. Laughton July 12, 80
C.B.W.R.Co. Ex. "H"
C.B.W.R.Co. Ex. "H"
Oliver H.
C.B.W.R.Co. Ex. "H"
Hd Abraham May
State
C.B.W.R.Co. Ex. "H"
C.B.W.R.Co. Ex. "H"
C.B.W.R.Co. Ex. "H"
State
State
C.B.W.R.Co. Ex. "H"
C.B.W.R.Co. Ex. "H"
P.R.E. Wm. O. Paxton Jan. 3, 85
C.B.W.R.Co. Ex. "H"
C.B.W.R.Co. Ex. "H"
Henry
C.B.W.R.Co. Ex. "H"
C.B.W.R.Co. Ex. "H"
Joshua Brown
C.B.W.R.Co.
Richard C.
C.B.W.R.Co.